You've Got Dissent!

Chinese Dissident Use of the Internet and Beijing's Counter-Strategies

Michael Chase • *James Mulvenon*

RAND

**National Security Research Division
Center for Asia Pacific Policy**

The research described in this report was conducted through the International Security and Defense Policy Center (ISDPC) of RAND's National Security Research Division (NSRD).

Library of Congress Cataloging-in-Publication Data

Chase, Michael S.
    You've got dissent! : Chinese dissident use of the Internet and Beijing's counter-strategies / Michael S. Chase, James C. Mulvenon.
      p. cm.
    "MR-1543."
    Includes bibliographical references.
    ISBN 0-8330-3179-1
    1. Internet—Government policy—China. 2. Internet—China. 3. Dissenters—China. 4. Civil rights—China. 5. China—Politics and government—1976– I. Mulvenon, James C., 1970– II. Title.

JQ1509.5.A8 C48 2002
320.951'0285'4678—dc21

                                    2002069720

RAND is a nonprofit institution that helps improve policy and decisionmaking through research and analysis. RAND® is a registered trademark. RAND's publications do not necessarily reflect the opinions or policies of its research sponsors.

Cover design by Stephen Bloodsworth

© Copyright 2002 RAND

Published 2002 by RAND
1700 Main Street, P.O. Box 2138, Santa Monica, CA 90407-2138
1200 South Hayes Street, Arlington, VA 22202-5050
201 North Craig Street, Suite 102, Pittsburgh, PA 15213-1516
RAND URL: http://www.rand.org/
To order RAND documents or to obtain additional information, contact Distribution Services: Telephone: (310) 451-7002; Fax: (310) 451-6915; Email: order@rand.org

Contrasting visions of the political consequences of information technology:

> The telescreen received and transmitted simultaneously. Any sound that Winston made, above the level of a very low whisper, would be picked up by it; moreover, so long as he remained within the field of vision which the metal plaque commanded, he could be seen as well as heard. There was of course no way of knowing whether you were being watched at any given moment. How often, or on what system, the Thought Police plugged in on any individual wire was guesswork. It was even conceivable that they watched everybody all the time. But at any rate they could plug in your wire whenever they wanted to. You had to live—did live, from habit that became instinct—in the assumption that every sound you made was overheard, and, except in darkness, every movement scrutinized.

> —George Orwell, *1984*

> Governments of the Industrial World, you weary giants of flesh and steel, I come from Cyberspace, the new home of Mind. On behalf of the future, I ask you of the past to leave us alone. You are not welcome among us. You have no sovereignty where we gather.

> We have no elected government, nor are we likely to have one, so I address you with no greater authority than that with which liberty itself always speaks. I declare the global social space we are building to be naturally independent of the tyrannies you seek to impose on us. You have no moral right to rule us nor do you possess any methods of enforcement we have true reason to fear.

> —John Perry Barlow, "A Declaration of the Independence of Cyberspace"

This report analyzes political use of the Internet by Chinese dissidents and the Chinese government's strategies to counter this activity. For this report, a broad definition of the term *dissident* is employed that includes not only political dissidents active in the People's Republic of China (PRC), but also activists residing overseas, members of the Falungong group, Tibetan exiles, and others who use the Internet for purposes considered subversive by Beijing. The report is based on interviews with numerous government officials in Washington, D.C., and Beijing, as well as discussions with dissidents, Falungong members, human-rights advocates, and academics based in China and North America. The conclusions are also informed by a comprehensive review of the growing academic literature on the political impact of the Internet in China and other authoritarian countries, as well as relevant Western and Chinese media reports. The authors conducted field research in several cities in China and performed extensive data-gathering on the Internet, including visits to hundreds of web sites, chat rooms, and bulletin board sites. The research reported here was initiated in early 2000; the report was revised and updated in January 2002.

This study was conducted in the International Security and Defense Policy Center of RAND's National Security Research Division (NSRD). NSRD conducts research and analysis for the Office of the Secretary of Defense, the Joint Staff, the Unified Commands, the defense agencies, the Department of the Navy, the U.S. intelligence community, allied foreign governments, and foundations.

Comments are welcome; they should be directed to the project's principal investigator, James Mulvenon; the director of RAND's International Security and Defense Policy Center, Stuart Johnson; or the director of RAND's Center for Asia Pacific Policy, Nina Hachigian.

James Mulvenon
RAND
1200 South Hayes Street
Arlington, VA 22202-5050
(703) 413-1100 x5225
E-mail: mulvenon@rand.org

Stuart Johnson
Director, ISDPC
RAND
1200 South Hayes Street
Arlington, VA 22202-5050
(703) 413-1100 x5470
E-mail: Stuart_Johnson@rand.org

Nina Hachigian
Director, Center for Asia Pacific Policy
RAND
1700 Main Street, P.O. Box 2138
Santa Monica, CA 90407-2138
(310) 393-0411 x6030
E-mail: Nina_Hachigian@rand.org

CONTENTS

FIGURES AND TABLES

Figures

Tables

In authoritarian societies from Saudi Arabia, to Cuba, to Myanmar, to the People's Republic of China (PRC), dissidents are using the Internet to organize and communicate with each other, to access banned information, and to draw support from a global network of activists and nongovernmental organizations (NGOs). At the same time, the governments of these countries are struggling to prevent activists from using the Internet to erode government controls over the flow of information and to promote political or social agendas that the regimes find threatening. This gives rise to a series of questions about the political impact of the Internet in authoritarian societies: Does the Internet provide dissidents with potent new tools that they can use to promote their causes, break through the barriers of censorship, and perhaps ultimately undermine the power and authority of nondemocratic regimes? Or, on the contrary, is it more likely that those authoritarian governments will use the Internet as another instrument to repress dissent, silence their critics, and strengthen their own power?

This report addresses the use of the Internet by Chinese dissidents, Falungong practitioners, Tibetan activists, and other groups and individuals in the PRC and abroad who are regarded as subversive by the authorities in China. It also examines the counterstrategies that Beijing has employed in its attempts to prevent or minimize the political impact of Chinese-dissident use of the Internet.

The arrival of the Internet has altered the dynamic between the Beijing regime and the dissident community. For the state, the political use of the Internet further degrades the Chinese Communist Party's

(CCP's) ability to control the flow of information it deems politically sensitive or subversive into China and within China. The party, however, still uses Leninist methods to crush potential organized opposition, and as a result, no organization with the capacity to challenge the CCP's monopoly on political power presently exists in China.

For dissidents, students, and members of groups such as Falungong, the Internet—especially through its two-way communication capabilities, e.g., e-mail and bulletin board sites (BBS)—permits the global dissemination of information for communication, coordination, and organization with greater ease and rapidity than ever before. Moreover, it allows these activities to take place in some instances without attracting the attention of the authorities, as exemplified by the unexpected appearance of an estimated 10,000 to 15,000 members of Falungong outside Zhongnanhai, the Chinese central leadership compound, in April 1999.

The capability of even one-way Internet communication—particularly e-mail "spamming"—enables the dissident community to transmit uncensored information to an unprecedented number of people within China and to provide recipients with plausible deniability in that they can claim that they did not request the information. In part because of dissident countermeasures (such as the use of different originating e-mail addresses for each message), the PRC is unable to stop these attempts to "break the information blockade." There is a trend toward more groups and individuals becoming involved in activities of this type, which some have dubbed a form of "Internet guerilla warfare."

Small groups of activists, and even individuals, can use the Internet as a force multiplier to exercise influence disproportionate to their limited manpower and financial resources. At the same time, however, enhanced communication does not always further the dissident cause. In some cases, it serves as a potent new forum for discord and rivalry among various dissident factions.

In its counterstrategies, the PRC regime has made some use of high-tech solutions, including blocking of web sites, e-mail monitoring and filtering, denial, deception, disinformation, and even the hacking of dissident and Falungong web sites. There is some evidence that Beijing's technical countermeasures are becoming increasingly sophisticated. In addition, some nongovernmental groups have launched "vigilante hacks" against dissident web sites, which compounds the difficulty of determining the level of official government sponsorship for such activities. Beijing's approach, however, is predominantly "low-tech Leninist," employing traditional measures such as surveillance, informants, searches, confiscation of computer equipment, regulations, and physical shutdown of parts of the information infrastructure.

The regime understands implicitly that the center of gravity is not necessarily the information itself, but the organization of information and the use of information for political action. The strategy of the security apparatus is to create a climate that promotes self-censorship and self-deterrence. This is exemplified by the comments of a Public Security Bureau official: "People are used to being wary, and the general sense that you are under surveillance acts as a disincentive. The key to controlling the net in China is in managing people, and this is a process that begins the moment you purchase a modem."

The government's strategy is also aided by the current economic environment in China, which encourages the commercialization of the Internet, not its politicization. As one Internet executive put it, for Chinese and foreign companies, "the point is to make profits, not political statements."

On the whole, Beijing's countermeasures have been relatively successful to date. The current lack of credible challenges to the regime despite the introduction of massive amounts of modern telecommunications infrastructure, however, does not inexorably lead to the conclusion that the regime will continue to be immune from the forces unleashed by the increasingly unfettered flow of information within and across its borders. While Beijing has done a remarkable job thus far of finding effective counterstrategies to what it perceives as the potential negative effects of the information revolution, the scale of China's information-technology modernization would suggest that time is eventually on the side of the regime's opponents.

ACKNOWLEDGMENTS

The authors would like to thank the numerous U.S. and Chinese individuals interviewed in the course of this research. To promote candor and for reasons of privacy, all must remain anonymous. They are identified in this report only with general titles, such as "pro-democracy activist," "Falungong practitioner," "Western business-person," or "PRC official." Moreover, in preparing this study, we were sensitive to the concerns of our interlocutors that its publication might reveal information that could compromise their efforts or perhaps even endanger individuals who reside in China or who travel frequently to the PRC. To protect our sources and others, some material considered particularly sensitive by our contacts was omitted from this report.

We also wish to extend our thanks to Professor David M. Lampton of The Johns Hopkins University School of Advanced International Studies (SAIS) and our RAND colleagues Robert Anderson and Nina Hachigian for reviewing this report prior to publication. Any errors or omissions are, of course, entirely the responsibility of the authors.

ACRONYMS

ACR	access-control router
APEC	Asia-Pacific Economic Cooperation
APNIC	Asia-Pacific Network Information Center
BAIIT	Beijing Application Institute for Information Technology
BBS	bulletin board site
CASS	Chinese Academy of Social Sciences
CCP	Chinese Communist Party
CDJP	Chinese Democracy and Justice Party
CDP	China Democracy Party
CDU	China Development Union
CND	*China News Digest*
CNNIC	China Internet Network Information Center
HRIC	Human Rights in China
ICB	International Connection Bureau
ICT	International Campaign for Tibet
IP	Internet protocol
ISP	Internet service provider
IT	information technology

MPS	Ministry of Public Security
MSS	Ministry of State Security
NGO	nongovernmental organization
PGP	Pretty Good Privacy (encryption software)
PNTR	permanent normal trade relations
PRC	People's Republic of China
PSB	Public Security Bureau
SOE	state-owned enterprise
TYC	Tibetan Youth Congress
WTO	World Trade Organization

Chapter One

POLITICAL USE OF THE INTERNET IN CHINA

INTRODUCTION

This report addresses the use of the Internet by Chinese dissidents, members of Falungong,[1] Tibetan activists, and other groups and individuals in the People's Republic of China (PRC) and abroad who are regarded as subversive by the Chinese authorities. It also examines the counterstrategies that those authorities have employed in their efforts to prevent or minimize the impact of dissident use of the Internet. Finally it attempts to assess future trends in these areas.

The report is based on interviews with numerous government officials in Washington, D.C., and Beijing, as well as discussions with dissidents, Falungong members, human-rights advocates, and academics based in China and North America. The conclusions are also informed by a comprehensive review of the growing literature on the political impact of the Internet in China and other authoritarian countries,[2] as well as relevant Western and Chinese media reports. In

---

[1]Banned in China, Falungong combines meditation with certain quasi-spiritual beliefs. For an introduction to Falungong, see the group's web sites, listed in the Appendix.

[2]Recent publications that deal specifically or in passing with the political impact of the Internet in China include Shanthi Kalathil and Taylor C. Boas, "The Internet and State Control in Authoritarian Regimes: China, Cuba, and the Counterrevolution," Carnegie Endowment Working Papers, No. 21, July 2001; Eric Harwit and Duncan Clark, "Shaping the Internet in China: Evolution of Political Control over Network Infrastructure and Content," *Asian Survey*, Vol. 41, No. 3, May/June 2001; Nina Hachigian, "China's Cyber-Strategy," *Foreign Affairs*, Vol. 80, No. 2, March/April 2001, pp. 118–133; William Foster and Seymour E. Goodman, *The Diffusion of the Internet in China*, Center for Security and Cooperation, Stanford University, November 2000; and

addition, the research for this project included fieldwork in several cities in China and extensive data-gathering on the Internet, including the examination of hundreds of Chinese-language web sites, chat rooms, and bulletin board sites (BBS).

This report presents a case study that focuses on China, but it deals with a number of themes that are relevant to the analysis of the political impact of the Internet in authoritarian states and other non-democratic regimes. As such, its conclusions make a contribution to

---

Guobin Yang, "The Impact of the Internet on Civil Society in China: A Preliminary Assessment" (forthcoming). For a journalistic account of China's dissident diaspora that touches upon the use of the Internet by activists in exile, see Ian Buruma, *Bad Elements: Chinese Rebels from Los Angeles to Beijing*, Random House, 2001. A revealing, though by the author's own admission not particularly objective, look at China's Internet entrepreneurs that also contains some interesting observations on the social and political side effects of the information-technology revolution is provided by David Sheff, *China Dawn: The Story of a Technology and Business Revolution*, New York: Harper Business, 2002.

The Internet's political consequences are also addressed in the U.S. Department of State's annual reports on human rights in China, the most recent of which, *China Country Report on Human Rights Practices, 2001*, was issued in March 2002. Also available to the public are several reports prepared by the U.S. embassy in Beijing, including "PRC Web Forums on Mid Air Collision," April 2001; "Web Discussion Sample: True Democracy, Fake Democracy, or No Democracy?" undated; "China: Information Security," June 1999; "China's Internet Information Skirmish," January 2000; "Kids, Cadres, and 'Cultists' All Love It: Growing Influence of the Internet in China," March 2001; and "PRC Net Dreams: Is Control Possible?" September 1997.

Chinese researchers are also conducting work in related areas. For example, see Guo Liang and Bu Wei, *Hulianwang shiyong zhuangkuang ji yingxiang de diaocha baogao* [*Investigative Report on Internet Use and Its Impact*], Chinese Academy of Social Sciences, Center for Social Development, research supported by State Informatization Office, April 2001; this paper is also available in an abridged English version, "The Questionnaire and Responses to a Survey on Internet Usage and Impact in Beijing, Shanghai, Guangzhou, Chengdu, and Changsha," Chinese Academy of Social Sciences, Center for Social Development.

A number of studies of the political influence of the Internet in other countries have also been published recently. On Burmese dissident use of the Internet and the regime's response, see Tiffany Danitz and Warren P. Strobel, "Networking Dissent: Cyber Activists Use the Internet to Promote Democracy in Burma," in John Arquilla and David Ronfeldt (eds.), *Networks and Netwars*, Santa Monica, CA: RAND, 2001, pp. 129–170. On the information revolution, network forms of organization, and "social netwar" in Mexico, see David Ronfeldt, John Arquilla, Graham E. Fuller, and Melissa Fuller, *The Zapatista Social Netwar in Mexico*, Santa Monica, CA: RAND, 1998. For a more general overview of the economic and political impact of the Internet and e-commerce at the local level in several developing nations, see Booz Allen & Hamilton, *E-Commerce at the Grass Roots: Implications of a "Wired" Citizenry in Developing Nations*, prepared for the National Intelligence Council, June 30, 2000.

the development of what is gradually becoming a substantial body of comparative literature on the topic of dissent and political activism in the digital age. The debate on this topic tends to center on the following questions: Does the Internet provide dissidents with potent new tools that they can use to promote their causes, break through the barriers of censorship, and perhaps ultimately undermine the power and authority of nondemocratic regimes? Or, on the contrary, is it more likely that those authoritarian governments will use the Internet as another instrument to repress dissent, silence their critics, and strengthen their own power?

These questions, though often presented in somewhat oversimplified form, are relevant in dozens of countries around the world. From Saudi Arabia, to Cuba, to Myanmar, to China, dissidents are using the Internet to organize and communicate with each other, to access banned information, and to draw support from a global network of activists and nongovernmental organizations (NGOs). At the same time, the governments of these countries are struggling to prevent activists from using the Internet to erode government controls over the flow of information and to promote political or social agendas that the regimes find threatening. These authoritarian regimes employ a variety of countermeasures, some of which rely on the latest advances in communications technology and some of which are much more traditional.

In most cases, these cat-and-mouse games between governments and dissidents have produced no clear winners, and as yet, there are no definitive answers to the fundamental questions raised above. The principal finding of this report, however, is that the arrival of the Internet in China has altered the dynamic between the Beijing regime and the dissident community. It has enabled local dissident efforts to become issues of national and even global discussion. It has allowed dissidents on the mainland to communicate with each other and with their counterparts in the exile dissident community, as well as with other overseas supporters, with greater ease and rapidity than ever before. In response, the Chinese government has employed a mixture of low-tech and high-tech countermeasures, with a degree of success thus far that has confounded many observers.

THE STATE OF THE INTERNET IN CHINA

The Internet has been at the forefront of the information revolution in China. While penetration of the Internet is still fairly insignificant when measured in either relative or absolute terms, growth in the number of users since 1995 has been virtually exponential and is expected to increase at geometric rates for the indefinite future. The number of Internet users in China had reached 33.7 million by January 2002, up from only 2.1 million in January 1999, according to the China Internet Network Information Center (CNNIC) (see Table 1).[3]

China's international connectivity and the number of computers in the PRC with Internet access are also increasing at rapid rates. International data bandwidth has grown dramatically in recent years, increasing almost tenfold between January 2000 and July 2001, and more than doubling between July 2001 and January 2002 (see Table 2), while the number of computers connected to the Internet, only around 750,000 in January 1999, had surpassed 12,500,000 by January 2002 (see Table 3).[4]

---

[3]China Internet Network Information Center, *"Zhongguo hulian wangluo fazhan zhuangkuang tongji baogao"* [Survey Statistical Report on the Development of the Chinese Internet], January 2002, p. 5. See also "Semiannual Survey Report on the Development of China's Internet," July 2001, and other previous CNNIC reports (listed in the References). The CNNIC, located in Beijing's Haidian district, was established in June 1997 and operates under the leadership of the Ministry of Information Industry and the Chinese Academy of Sciences. Its responsibilities include registering domain names, distributing IP (Internet protocol) addresses, and conducting statistical surveys on the development of the Internet in China. The CNNIC has released nine of these survey reports. The first two were issued in October 1997 and July 1998. Since then, the CNNIC has issued reports twice annually, every January and July. The reports are available online (all but the first two reports are available in English translation) at CNNIC's website: www.cnnic. gov.cn/e-index.shtml. CNNIC reports also provide information on total bandwidth, Internet user demographics, number and geographic distribution of domain names and web sites, access locations and expenditures, and user views regarding online advertising and e-commerce. It should be noted that industry experts have questioned the methodology employed by the CNNIC to count Internet users and to tabulate some of the other statistics in its reports. It is also worth noting that account-sharing is another factor that complicates attempts to estimate accurately the number of people in China who are "wired."

[4]Ibid., pp. 5–7. More than half of all Chinese Internet users access the Internet at their homes or offices, according to the July 2001 report, while around 18 percent access the Internet at school, and 15 percent access it at Internet cafes. January 2000 statistics are from CNNIC, "Semi-Annual Survey Report on Internet Development in China," January 2000.

Table 1

Growth of Internet Use in China

Date	Number of Users
October 1997	620,000
July 1998	1,175,000
January 1999	2,100,000
July 1999	4,000,000
January 2000	8,900,000
July 2000	16,900,000
January 2001	22,500,000
July 2001	26,500,000
January 2002	33,700,000

Source: China Internet Network Information Center (CNNIC).

Table 2

Total Bandwidth of China's International Connections

Date	Total Bandwidth (megabits/second)
October 1997	25
July 1998	85
January 1999	134
July 1999	241
January 2000	351
July 2000	1,234
January 2001	2,799
July 2001	3,257
January 2002	7,598

Source: China Internet Network Information Center (CNNIC).

At the same time, however, there are a number of important distortions in the Internet modernization process that impede the effectiveness of the Internet as a tool for political expression. First, as revealed by the latest official statistics, a pronounced "digital divide" exists in China. Internet usage continues to be dominated by an extremely narrow sliver of the national demographic, primarily young, highly educated, urban men. According to the most recently released official statistics:

Table 3

**Number of Computers Connected to the Internet
in China**

Date	Number of Computers
October 1997	299,000
July 1998	542,000
January 1999	747,000
July 1999	1,460,000
January 2000	3,500,000
July 2000	6,500,000
January 2001	8,920,000
July 2001	10,020,000
January 2002	12,500,000

Source: China Internet Network Information Center
(CNNIC).

- More than two-thirds of all Internet users in China are 30 years of age or younger, and nearly 80 percent are under 35.

- Approximately 90 percent of the Internet users have attained at least a high school diploma, and more than 60 percent have attended college.

- Users are heavily concentrated in prosperous provinces and municipalities in China's economically dynamic coastal areas, primarily Beijing, Shanghai, Jiangsu, Zhejiang, and Guangdong.

- Of the estimated 33.7 million Internet users in China, only 1.9 percent are peasants or farmers.

- Roughly 60 percent of the Chinese Internet users are male, although the proportion of Chinese women using the Internet has grown steadily over the past several years.[5]

While the current Internet-using cohort is likely to be most amenable to pro-reform political messages from abroad, the failure of a similarly structured protest movement in 1989 highlights the importance of a grass-roots, inclusive base for political change. Also, recent

---

[5]Ibid., pp. 5–10. The latest CNNIC report also shows that northeastern rustbelt provinces and inland areas host only a tiny percentage of Chinese web sites.

trends in China indicate that this cohort is becoming highly national-istic in outlook, and anecdotal evidence suggests that its members have ambivalent attitudes toward the United States. Moreover, only slightly more than 20 percent of the information viewed by Chinese Internet users is in languages other than Chinese, and many users identify high prices[6] and slow access speed as serious, continuing problems.[7] The second developmental distortion centers on the shifting government regulatory environment regarding the Internet in China, which appears to be creating a situation in which coopera-tion or even partnership with government organs is the only effective strategy for profitable investment. The close ties between govern-ment and the commercial Internet sector, as well as a set of harsh edicts about improper Internet use, have led Internet service pro-viders (ISPs) to implement policies of self-censorship and have de-terred Internet users from fully exploiting the political potential of the medium.

THE STATE OF UNSANCTIONED NGOs INSIDE CHINA

The existence of the Internet, of course, is not inherently threatening to the political position of the Chinese Communist Party (CCP). But when the Internet is used by unsanctioned organizations inside China for "subversive" purposes, the regime takes notice. This sec-tion assesses the state of unsanctioned NGOs in China and provides a brief introduction to the major groups that Beijing considers sub-versive users of the Internet and related communications tech-nologies.

The CCP has never tolerated the formation of organizations that at-tempt to exist outside of its control, and over the past several years, Beijing has effectively carried out Jiang Zemin's December 1998 pledge to "resolutely nip in the bud" all attempts to form unsanc-tioned political organizations in China.[8] Although Chinese authori-

---

[6]For an extensive discussion of the cost issue, see Eric Harwit and Duncan Clark, "Shaping the Internet in China: Evolution of Political Control over Network Infra-structure and Content," pp. 390–391.

[7]CNNIC Report, January 2002, pp. 8–9.

[8]Jiang made this comment in a December 18, 1998, speech marking the 20th anniversary of the Third Plenum of the 11th CCP Central Committee. See "Jiang

ties have been unable to prevent sporadic demonstrations by Falungong adherents or to completely silence the surprisingly resilient China Democracy Party (CDP), there is at present no political, social, or religious organization in China with the capability to challenge the CCP's monopoly of political power.

Over the past several years, the Falungong meditation sect and the CDP were the two unsanctioned groups that Beijing perceived as posing the most serious challenges to its authority. But even as it focused its efforts on suppressing these two groups, especially Falungong, the regime was bedeviled by a host of other organizations, including many based outside of China. The most prominent among these groups are the Tibetan exile community and its global network of supporters, and the dissident diaspora in North America and Europe. Even individual activists, such as Frank Siqing Lu, the director of the Hong Kong Information Center, which gathers information on dissident arrests and worker demonstrations and faxes handwritten press releases to international media organizations, caused Beijing great concern during this period of time and continue to do so today.

Before discussing dissident use of the Internet, we present brief sketches of three of the principal unsanctioned organizations that are discussed throughout this report: Falungong, the CDP, and the Tibetan exile movement.[9]

Falungong

Falungong was founded by Li Hongzhi, a government clerk, in 1992. At first glance, the organization appears similar to hundreds of groups in China that practice *qigong*, a traditional system of exercise

---

Stresses Stability, Unity," *Xinhua*, December 18, 1998, in FBIS, December 18, 1998. At a national law-enforcement conference three days later, Jiang declared, "Stability is the prerequisite for reform and development." He warned, "Without stability, we would achieve nothing and forfeit what we have achieved." Jiang told the assembled members of the police and law-enforcement apparatus that maintaining stability was thus of the "utmost importance." See "Jiang, Zhu Speak at Politics and Law Conference," *Xinhua*, December 23, 1998, in FBIS, December 24, 1998.

[9]We also mention the use of the Internet by other dissident groups and individual activists, and several are discussed in greater detail in other sections of the report. Because they are not the most prominent actors, however, these groups and individuals are not examined at length here.

and healing. What distinguishes Falungong from the other groups, however, is its founder, who claimed to have received knowledge of a spiritual system from "20 masters" who sought him out as a child to educate him in its ways. To traditional *qigong*, which holds that certain physical exercises can channel human energies in healing ways, Li added his own supernaturally received wisdom. The tenets included teaching far more potent healing exercises than are typically associated with *qigong*, including the healing of terminally ill persons; various cosmological and moral precepts; and the teaching that very accomplished practitioners of Falungong can attain "supernormal" powers such as teleportation and the ability to fly. By the mid-1990s, Li had gathered notes on his teachings into pamphlets and books that he sold at his classes and lectures. After clashes with the authorities, Li left China with his wife and daughter to live in the United States. Various estimates of the number of Falungong practitioners range from the understated Chinese government estimate of 10 million members to the group's own July 1999 estimate of 100 million members in more than 30 countries.[10] One independent observer placed the number at around 40 million in 1999.[11]

Falungong came to national and global attention on April 25, 1999, when between 10,000 and 15,000 Falungong practitioners gathered outside the Zhongnanhai central leadership compound in Beijing. The protest was reportedly prompted by reports of violence inflicted on Falungong practitioners by Chinese police in Tianjin two days earlier, as well as an official ban on the publishing of Falungong materials. The protesters arrived to petition the National People's Congress and the party leadership, with the goal of obtaining official registration with the Ministry of Civil Affairs, the Religious Affairs Bureau, or the China Buddhist Association. The protesters did not give the police any excuse for high-handed suppression. No banners were displayed, no slogans were chanted, and the crowd of largely elderly and female practitioners shunned contact with the foreign media. According to one account, the protesters sat in complete silence, ar-

---

[10]For the Falungong estimate, see "A Report on Extensive and Severe Human Rights Violations in the Suppression of Falungong in the People's Republic of China, 1999–2000," compiled by the group.

[11]"Falungong and the Internet: A Marriage Made in Web Heaven," VirtualChina.com, July 30, 1999.

ranged in neat rows, for 18 hours. They cooperated with police, dispersed peacefully at nightfall, and even collected their own garbage.

By all accounts, the Chinese government had no foreknowledge of the demonstration, which was coordinated through the use of the Internet and wireless telephones.[12] In public, the government initially responded by adopting a moderate position, announcing that the protest would not result in punishment of those involved.[13] Behind the scenes, however, the CCP was preparing its counterattack. First, the leadership of the Ministry of Public Security (MPS) was allegedly criticized for not providing any warning of the protest to the leadership or preempting the march itself, and some of the ministry's leaders were purged. After two months of investigation, which revealed a surprising number of practitioners among the ranks of military officers and ministry officials, a nationwide crackdown on Falungong was initiated.[14] The crackdown combined a counterpropaganda campaign with an aggressive series of arrests and detentions.[15] On July 20, 1999, hundreds of key members of Falungong were reportedly arrested in the middle of the night. On July 21, the group was banned in China for allegedly spreading "superstitious, evil thinking." President Jiang Zemin allegedly ordered the ban on the sect at an emergency Politburo meeting, asserting that Falungong was a destabilizing force. The Public Security Bureau issued a notice forbidding engaging in protests, spreading rumors, putting up notices or banners, or disseminating publications relating to Falun-

---

[12]See Kevin Platt, "China's 'Cybercops' Clamp Down; Beijing Sees Growing Web Use as Threat, But It Had a Victory Nov. 9 in Connection with Four Convictions," *Christian Science Monitor*, November 17, 1999, p. 6; and Willy Wo-Lap Lam, "Falungong Protest Shocks Party Leadership," *South China Morning Post*, April 28, 1999, p. 17.

[13]"Talks Given by Officials of the State Council and the Chinese Communist Party Central Letters and Visit Bureaus," *Xinhua*, June 14, 1999, in FBIS, June 14, 1999.

[14]John Pomfret, "China Sect Penetrated Military and Police: Security Infiltration Spurred Crackdown," *Washington Post*, August 7, 1999.

[15]The length of time between the demonstration and the crackdown can be explained by the general lack of knowledge about Falungong and its activities among members of the leadership and the security apparatus. Jiang Zemin himself has reportedly told foreign visitors that he had never heard of the group before the gathering. Various government organs needed to carry out detailed investigations of the movement after the demonstration, to provide data for the deliberations at higher levels.

gong. Other circulars prohibited party members from joining the sect. A week later, an arrest warrant was issued for Li Hongzhi, who by then was already a permanent resident of the United States.

The Chinese security apparatus' tried-and-true strategy of leadership decapitation and intimidation of the rank-and-file, which has been highly successful in past crackdowns, has not worked completely with Falungong. At the time of the official ban, the regime's attitude was summarized by a high-ranking official: "We will cope with any kind of reaction. The majority will have left the organization now that it's illegal." In fact, however, the government appears stymied by the cellular organizational structure of the group. To the alleged consternation of the leadership, practitioners continued to protest and file petitions throughout 1999 and 2000. In October 1999, for instance, Falungong held a clandestine press conference in Beijing, attended by foreign journalists. Shortly thereafter, the PRC government raised the ante by declaring Falungong to be an "evil cult," permitting the levying of harsher legal penalties against the group. Official state media outlets have churned out countless print, radio, and television stories detailing Falungong's "crimes," including the leaking of state secrets,[16] the masterminding of 307 demonstrations, and the deaths of 743 people.[17] In addition, millions of Falungong books, audiotapes, and videotapes have been publicly destroyed.[18] The campaign also has an international dimension: Chinese embassies abroad have been ordered to disseminate anti-Falungong material in multiple languages.

Recently, the frequency of demonstrations has declined, as large numbers of practitioners have been imprisoned, sent to labor camps, or driven underground. In addition, by some accounts, the fiery suicide of several Falungong members in Tiananmen Square on January 23, 2001, and the ensuing campaign to vilify the group in the official media, which featured graphic images of a young girl who died from burns suffered during the incident, has turned popular opinion

---

[16]"Police Claim Falungong Followers Leaked State Secrets," *Xinhua,* October 25, 1999, in FBIS, October 25, 1999.

[17]See Pomfret, "China Sect Penetrated Military and Police."

[18]"Circular Issued on Destroying Falungong Publications," *Xinhua,* July 28, 1999, in FBIS, July 28, 1999.

decidedly against Falungong on the mainland and perhaps even in some overseas Chinese communities.

The China Democracy Party

On June 24, 1998, emboldened by the relatively relaxed political climate of the 1997–1998 "Beijing Spring," three dissidents in Hangzhou, Zhejiang Province, announced the establishment of the CDP.[19] The dissidents timed their application for official recognition to coincide with President Clinton's visit to China in June 1998, apparently confident that Beijing would not order local authorities to act against them during the Sino-U.S. summit.[20] Within a few months, dissidents in locations throughout China had established regional branches of the nascent opposition party. In November, members of the CDP announced plans to convene a national party congress and applied to the State Council for permission to form a "national preparatory committee."

Increasingly concerned about social and political stability at home, and with the perceived utility of tolerating dissident activity declining as the overall state of U.S.-China relations deteriorated, Beijing decided to launch a crackdown against the CDP. In late December 1998, the CDP's three most prominent members, Wang Youcai, Xu Wenli, and Qin Yongmin, were sentenced to long prison terms on charges of "endangering state security." At least two dozen more CDP members have since been imprisoned, and many others are being held in detention.

Although the CDP called for wide-ranging reforms, it adopted a relatively moderate approach by organizing openly and applying to the authorities for permission to register legally. The CDP even declared that, while it opposed China's authoritarian political system, it did not oppose the CCP itself and would be willing to work with the CCP

---

[19]The group was led by Wang Youcai, a student leader of the 1989 Tiananmen Square demonstrations who was named on Beijing's "most wanted" list following the June 4 crackdown.

[20]See "Open Declaration on the Establishment of the China Democracy Party Zhejiang Preparatory Committee" (*Zhongguo minzhudang Zhejiang choubei weiyuanhui chengli gongkai xuanyan*), available on the CDP web site or by request from the authors.

to achieve political reform.[21] Nevertheless, Beijing regarded the dissidents' attempt to organize a nationwide opposition party as an intolerable assault on the "Four Cardinal Principles."[22] Despite Beijing's crackdown, a number of CDP members remain active on the mainland. The CDP has also developed an organization-in-exile in the United States.[23]

The Tibetan Exile Community

Between late 1949 and September 1951, Chinese troops invaded Tibet, eventually conquering the entire country, which Beijing claimed had always been part of the pre-modern Chinese empire. For a time, the Dalai Lama and the Tibetan authorities attempted to co-exist with their new rulers. By March 1959, however, resistance to Chinese rule had grown into a national uprising. The rebellion was quickly and ruthlessly crushed by Chinese forces, resulting in the deaths of thousands of Tibetans. The Dalai Lama and some 80,000 Tibetan refugees fled the capital city of Lhasa, taking asylum in the Indian city of Dharmsala, which is now the home of the Tibetan government-in-exile. According to statistics provided by Tibetan authorities, there are currently more than 130,000 Tibetan refugees around the world.

---

[21]See, for example, "*Zhongguo minzhudang zhengzhi gangling*" [Political Program of the China Democracy Party], available on the CDP web site or by request from the authors. In its political program, the CDP states that it "does not oppose the Communist Party, it only opposes the authoritarian system, and it seeks with the CCP a goal where both sides win" (*bu fandui gongchandang, zhi fandui zhuanzhi zhidu, zhuiqiu yu zhonggong shuangying de mubiao*). See also "Pro-Democracy Activist Inter-viewed on Party Formation," *Mainichi Shimbun*, September 13, 1998, in FBIS, September 13, 1998. Xie Wanjun, then a CDP member in Shandong Province, stated that the CCP would continue to be the ruling party during the process of political reform and pledged, "If our request to form a party is approved, we will act within the scope of the current constitution." Xie added, "We can form the party peacefully after holding discussions with the CCP without confrontation."

[22]The Four Cardinal Principles, sometimes referred to as the Four Basic Rules or the Four Upholds, state that China must uphold (1) the leadership of the CCP, (2) Marxism-Leninism-Mao Zedong thought, (3) the people's democratic dictatorship, and (4) the path of socialism.

[23]Several interviewees, including exiled dissidents and Western NGO researchers, noted that the actual size of the U.S.-based component of the CDP is unclear.

From their base in India, the Dalai Lama and his people have continued to press for greater autonomy for Tibet. Rather than negotiating with the government-in-exile, Beijing has implemented policies of ethnic Chinese resettlement and cultural and religious repression of Tibetans. Activists assert that as a result of these measures, an unknown number of Tibetans have been killed since the original annexation, and more than 6,000 monasteries have been destroyed. Han Chinese now outnumber Tibetans in Tibet by almost 1.5 million, and parts of historic Tibet have been incorporated into five neighboring Chinese provinces. Recent attempts to foster communication between the Tibetan government-in-exile and Beijing have foundered over disagreements about the choice of the current reincarnation of the Panchen Lama, one of the most important religious figures in Tibetan Buddhism; disputes over the size of Tibet; and Beijing's mistrust of and vituperative verbal attacks against the Dalai Lama.

Activities committed to winning increased autonomy for Tibet are supported principally by the Tibetan government-in-exile, composed of a variety of elected and civil service institutions. In addition, there are six central Tibetan NGOs: the Tibetan Youth Congress, the Tibetan Women's Association, Cholkha-sum, the Ngari Association, Ghu-chu-sum (a support organization for former Tibetan political prisoners), and the United Association. These NGOs are joined by an extensive international network of Tibetan advocacy organizations, including Students for Free Tibet, the International Campaign for Tibet, the International Tibet Independence Movement, the Tibet Support Group, the Tibet Fund, the Milarepa Fund (sponsors of the annual Tibetan Freedom Concerts), the Committee of 100 for Tibet, Tibet Environmental Watch, the Tibetan Plateau Project, the International Committee of Lawyers for Tibet, Tibet House, and the Tibetan Center for Human Rights and Democracy.

USE OF THE INTERNET

We next examine the use of the Internet by mainland and exile dissidents, Falungong practitioners, members of the Tibetan diaspora, and other activists for both two-way and one-way communication.

Two-Way Communication

For dissidents, students, and members of groups such as Falungong, the Internet permits the global dissemination of information, especially through two-way communication such as e-mail and BBS, for communication, coordination, and organization with greater ease and rapidity than ever before. Importantly, it allows the activists to pursue their activities in some instances without attracting the attention of the authorities, as exemplified by the unexpected Falungong demonstration outside the central leadership compound in April 1999. Dissidents, Falungong adherents, Tibetan exiles, and Chinese university students use a variety of means of two-way Internet communication, including e-mail, web-based petitions, BBS, and chat rooms to coordinate, organize, motivate, and transmit information regarded by Beijing as politically sensitive or "subversive."

Two-Way E-mail Communication and Coordination. E-mail is an especially important tool for two-way communication and coordination among dissidents, Falungong practitioners, and Tibetan exiles. A variety of evidence indicates that mainland dissidents regularly use e-mail, as well as Internet chat rooms and bulletin boards, to communicate and coordinate with each other and with members of the exile dissident community.[24] Soon after Internet service became available to the public in China, several prominent dissidents began to use e-mail to organize political discussion salons and seminars.[25] More recently, two-way communication via e-mail played an important role in the organization and development of the CDP. Several CDP members assert that use of e-mail and the Internet was critical to the formation of the party and allowed its membership to expand from about 12 activists in one region to more than 200 in provinces and municipalities throughout China in only four months.[26] The Zhejiang branch of the CDP reportedly became a

---

[24]See for example, "Internet Allows Chinese Dissidents to Network," www.nando.net, June 2, 1998. Ren Wanding says that he uses e-mail on an almost daily basis to get information that is not available in the official media, to keep track of news regarding other mainland dissidents, and to contact members of the overseas democracy movement.

[25]Interview, U.S. government executive branch official, May 2000.

[26]Maggie Farley, "Hactivists Besiege China," *Los Angeles Times*, January 4, 1999. See also Jasper Becker, "Review of Dissidents, Human Rights Issues," *South China*

particularly important and influential component of the party, in part because many dissidents in the region owned computers and had e-mail access.[27] Over the past several years, mainland dissidents have also almost certainly used e-mail to coordinate open letters and petitions, many of which were signed by 100 or more dissidents.[28] In addition, in early 1999, mainland and overseas activists used e-mail to coordinate and publicize an abortive attempt to form the China Labor Party.[29]

The use of e-mail is equally important to exile dissident groups such as the Chinese Democracy and Justice Party (CDJP). According to Shi Lei, director of the CDJP's Internet division, "The use of the Internet and e-mail to transmit information about the democracy movement has been the most effective method of communication for the CDJP since its founding."[30] Shi says that overcoming the countermeasures employed by the Chinese authorities is a constant struggle. "The CCP continuously tries to blockade us," he says, "and we never stop looking for new ways to break the blockade."[31]

Morning Post, January 12, 1999. Beijing CDP member and longtime dissident Gao Hongming told Becker, "It is the first time we attracted so many people from all over the country; it shows what can be done."

[27]"Police Arrest Dissidents to Prevent Seminar Opening," Hong Kong Information Center of Human Rights and Democratic Movement in China, March 14, 1999, in FBIS, March 14, 1999.

[28]In late December 1998, for example, 274 dissidents from 20 provinces signed an open letter demanding that authorities release Hunan dissident Zhang Shanguang, who was sentenced to 10 years' imprisonment for discussing rural unrest on Radio Free Asia. It seems very likely that the signatories relied to a great extent on e-mail to distribute and coordinate the appeal.

[29]"New Party for Workers to Seek Registry," Hong Kong *Agence France Presse*, January 2, 1999, in FBIS, January 2, 1999. A Beijing dissident using the alias "Li Yongming" and a U.S.-based exile announced the attempted formation of the party to the Hong Kong media organization via e-mail. They announced that the party's goals included combating official corruption and said it would draw its supporters in part from the ranks of laid-off workers, a prospect Beijing must have found particularly worrisome given widespread discontent among the growing numbers of unemployed workers and the increasing frequency of demonstrations.

[30]Shi Lei, "*Xinxi bailinqiang: tupo zhonggong wangluo dianzi youjian fengsuo (zhiyi)*" [The Information Berlin Wall: Breaking the Chinese Communist Party's Net and E-mail Blockade (part one)], available in Chinese only on the website of the Home for Global Internet Freedom at http://internetfreedom.org/gb/articles/994.html.

[31]Ibid.

For Falungong, e-mail is perhaps even more vital. When Li Hongzhi moved to the United States, Falungong set up e-mail lists to facilitate communication between followers in the United States, and between followers in the United States and China. The movement reportedly used e-mail to coordinate the April 1999 gathering outside the Zhongnanhai central leadership compound in Beijing.[32] Falungong also reportedly used e-mail to set up a secret press conference in Beijing to tell the world about police beatings of detained members.[33] Following Master Li's retreat from public life, all of his new articles have been disseminated via the group's primary e-mail list, Minghui Net. The importance of this list to the organization has become even greater in the aftermath of serious attacks against its bulletin boards and chat rooms, and repeated efforts to flood the Minghui Net mailbox have forced its administrators to adopt new methods. A July 27, 2000, message entitled "Notice to Overseas Practitioners from the Minghui Editors" announced that the old editorial mailbox (eng_editor@minghui.ca) was to be replaced by a new address (eng_article@minghui.org).[34] In the event that both mailboxes were unavailable, practitioners were instructed to send their articles "to the various local dafa associations and ask those in charge to submit them . . . to the Minghui editors." Practitioners were also encouraged to "periodically compress Minghui essays and documents into ZIP files and send them to the many readers in China and other regions where it is not convenient for them to access the Minghui site." It must be noted, however, that every submission to Minghui Net is retransmitted to the group at large. According to a July 14, 2000, posting entitled "On Important Matters, Practitioners Must Watch the Position of Minghui Net," users were told that their messages would be vetted by "a group of practitioners responsible for careful and detailed proofreading, editing, and approval."[35]

---

[32]Kevin Platt, "China's 'Cybercops' Clamp Down; Beijing Sees Growing Web Use as Threat, But It Had a Victory Nov. 9 in Connection with Four Convictions," *Christian Science Monitor*, November 17, 1999, p. 6.

[33]Ibid., p. 6; see also Erik Eckholm, "China Sect Members Covertly Meet Press and Ask World's Help," *New York Times*, October 29, 1999.

[34]See http://www.clearwisdom.net/eng/2000/July/27/EDITOR072700.html.

[35]See http://www.clearwisdom.net/eng/2000/July/16/AW071600_1.html.

Dissidents and human-rights activists also use the Internet and e-mail to transmit to the international media information about arrests, human-rights violations, and worker demonstrations on the mainland. Frank Siqing Lu's Hong Kong Information Center relies primarily on a beeper and callback system to gather information from informants in China[36] but has reportedly also made extensive use of e-mail for this purpose.[37] Indeed, Lu's e-mail address is listed on his web site, and he encourages visitors to the site to contact him with news of demonstrations by laid-off workers, retired workers, or farmers; the arrest or sentencing of individuals for engaging in political or labor-movement activities; infringements on religious freedom; banning of books or closure of magazines; and repression of Tibetans, Uyghurs, members of Zhonggong, or Falungong practitioners. Lu has even stated his hope that he might eventually be able to provide "spy-rate digital cameras" to mainland informants so that they can transmit pictures of worker demonstrations to him via e-mail.

Two-way e-mail communication is also an important channel for dissidents, Falungong members, Tibetan exiles, and various NGOs and human-rights advocacy groups to advance their interests with members of the U.S. Congress, congressional staff, and executive branch officials. Several interviewees in this study stated that e-mail has become the primary vehicle for such efforts to communicate with and energize U.S. government officials.[38] Although e-mail seemingly facilitates access to these officials, it may not always be an effective means of communication. Indeed, one interviewee opined that some Congressional staff are now deluged with so many e-mail

---

[36]For a detailed description (available only in Chinese) of this beeper and callback method of communication, see Lu's web site, which contains several pages of instructions explaining how to contact him from cities throughout China. Lu provides his beeper number, which ends with the digits 0604 to commemorate the June 4, 1989, Tiananmen crackdown, and states that no matter where a prospective informant is located, it is easy to reach him. Callers should simply dial his beeper number from a public telephone and he will call them back. The information on the beeper system is located at www.89-64.com/ct/191.html.

[37]Interviews, Western NGO employees, May 2000.

[38]Interviews, U.S. government executive and legislative branch officials, May–June 2000.

messages from activist groups on a daily basis that the messages have largely lost their effect.[39]

Interestingly, although many dissidents express concerns about the privacy and security of electronic communications, we found no indications that they make use of widely available encryption software, such as PGP (Pretty Good Privacy), to protect their e-mail communications from the Public Security Bureau and the Ministry of State Security (MSS). At present, like many other Chinese Internet users, most dissidents apparently fear that using encrypted e-mail would draw the attention of the security services. Recent postings on Chinese-language websites concerning the use of encryption suggest, however, that in the future, larger numbers of activists may turn to free, widely available encryption software products to enhance the security of their e-mail communications.[40]

Two-Way Communication Via Web-Based Petitions. Several groups of activists based in the United States and China have launched petition drives via the World Wide Web or supplemented traditional signature-gathering campaigns with an Internet component. To mark the 11th anniversary of the Tiananmen crackdown, Ding Zilin's Tiananmen Mothers Campaign and Human Rights in China (HRIC) established a web site that features a form-based electronic petition to Jiang Zemin and Li Peng.[41] The site, which is maintained in New York by HRIC, also urges visitors to become "virtual human-rights activists" by forwarding information to friends, listserves, and newsgroups, or by posting a link to the Tiananmen Mothers Campaign on

---

[39]Ibid.

[40]Examples include "*Yong jiami fangshi anquan shiyong dianzi youjian*" [Using Encryption for E-Mail Security], available at www.internetfreedom.org/gb/articles/987.html; "*Shenme shi PGP?*" [What is PGP?], November 12, 2001, available at http://www.internetfreedom.org/gb/articles/1034.html; and "*Youguan jiamifa, yinshenshu, ji yinxieshu ziyuan*" [Resources on Cryptography, Anonymity, and Steganography], November 12, 2002, available at http://www.internetfreedom.org/gb/articles/1035.html.

[41]The Tiananmen Mothers Campaign e-petition demands (1) the right to mourn in public for victims of the crackdown, (2) the right to receive aid from organizations outside of China, (3) cessation of the persecution of victims of the crackdown and their family members, (4) the release of all individuals still imprisoned on June 4–related charges, and (5) a public accounting for the killings. The site is located at www.fillthesquare.org.

their own web sites.[42] The site allows visitors to place a "virtual bouquet" of six white and four red roses in a cyberspace version of Tiananmen Square to mourn the victims of the June 4, 1989, crackdown. Organizers hope that the bouquet graphic will become a widely recognized political symbol that other activists can place on their web sites.[43]

In 1999, to commemorate the 10th anniversary of the Tiananmen Square crackdown, a group led by Wang Dan initiated a petition drive calling for the reversal of the official verdict on the student demonstrations. Approximately 20,000 visitors to Wang's web site, www.June4.org, signed the petition electronically, according to a press release from Wang's group. Although authorities in China attempted to block access to the site, more than 2,000 signatures came from Internet users who identified themselves as PRC residents.[44] Campaign organizers described the web site and e-petition as important supplements to traditional methods and a useful vehicle for advertising the campaign.[45] Chinese residents have even begun using the Web for domestic petition drives. An Jun, founder of an anticorruption newsletter in Henan Province, used the Internet as part of an effort to collect signatures on an anticorruption petition in mid-1999.[46]

Two-Way Communication via BBS and Chat Rooms. Chinese-language BBS such as the popular North America Free Talk Forum and the *People's Daily*'s Strong Country Forum (*Qiangguo luntan*) feature commentary on a variety of sensitive political, social, and economic topics from Internet users in the United States, Taiwan,

---

[42]Press release from HRIC, "Tiananmen Mothers Seed Global Support for Campaign to End Impunity: On-Line Petition Launched," June 2, 2000.

[43]Interviews, U.S.-based Chinese activists, June 2000.

[44]"Wang Dan Demands Change in China," press release of the Global Petition Campaign/June4.org. The electronic petition was also made available on other web sites, including those of Amnesty International and HRIC. Wang's group collected more than 150,000 signatures in all. Mainland Internet users also reportedly marked the 10th anniversary of the June 4, 1989, crackdown by posting the message "6+4=10" in chat rooms.

[45]Interviews, Chinese pro-democracy activists based in the United States and Canada, June 2000.

[46]"Dissidents to Start Signature Campaign," Hong Kong Information Center, May 20, 1999, in FBIS, May 20, 1999.

China, and elsewhere.[47] The number of postings to the forums has tended to peak around the time of important events. For example, our observations indicate that traffic on these sites increased around the time of the diplomatic standoff that ensued after the April 2001 collision of a U.S. reconnaissance aircraft and a Chinese fighter aircraft. In addition, in 2000, the number of postings on such sites surged around the times of the Taiwan presidential election, Chen Shui-bian's inauguration, the permanent normal trade relations (PNTR) vote in the U.S. Congress, and the anniversary of the June 4, 1989, Tiananmen Square crackdown.[48]

The sites that are hosted within the PRC are monitored, and censors often delete politically sensitive postings. According to a Chinese researcher, at least 1.5 percent of all postings to the popular Strong Country Forum web site, which was established by the CCP's main official newspaper soon after the May 1999 accidental bombing of the Chinese embassy in Belgrade, were censored in 2000.[49] The site's managers employ a three-part system to restrict content: First, rules listed on the web site forbid the posting of comments that call into question the validity of the Four Cardinal Principles or the party's policies, encouraging users to censor themselves. Second, the site reportedly uses software that screens postings for key words, such as the names of party leaders, and sends postings that contain those words to a webmaster for review. The third layer of censorship is the webmaster, who periodically deletes "problematic" postings that are not caught by the filtering software and thus appear, even if only temporarily, on the web site. The Chinese researcher also notes that the webmasters sometimes ban individual posters from contributing comments to the forum by blocking messages from their IP ad-

---

[47]In addition to messages having content of a political nature, there are also numerous messages advertising free e-mail services and "get paid while you surf" opportunities, as well as postings on other topics that lie beyond the scope of this report. Messages that use a variety of epithets to insult other users, known as "flames," are also frequently posted on some popular sites.

[48]Conclusions based on data gathered during April 2001 and between April 2000 and July 2000.

[49]Chinese researcher, January 2001. It should be noted that the methodology the researcher employed to generate this estimate yielded results that cannot be independently verified.

dresses.[50] In some cases, postings have even led to arrests. Fu Lijun, 37, an assistant professor at Xinxiang Medical College in Henan, was arrested in October 1999 for posting an article in a chat room detailing how Falungong could cure illness.[51] In December 2001, Wang Jinbo, a member of the CDP, was sentenced to four years in prison for posting on the Internet a message urging Beijing to reevaluate the 1989 Tiananmen movement.[52]

Still, not all comments on sensitive political topics are quickly expunged, and such missives have even appeared on some of the BBS managed by official Chinese media organs. After Chen Shui-bian's inauguration on May 21, 2000, for example, some Internet users suggested on the Strong Country Forum site that China should adopt a democratic system to promote reunification, while several others criticized CCP leaders for failing to deal more firmly with Taiwan.[53] Apparently overwhelmed by the deluge of postings, *People's Daily* censors were unable to immediately expunge all such messages.

Several dissident groups maintain their own BBS. The CDP, for example, established more than a dozen Chinese-language BBS in May 2000.[54] These include an organizational-development forum (*zuzhi fazhan luntan*), a 6-4 Tiananmen forum (*liu-si Tiananmen luntan*), a mainland democracy-movement forum (*dalu minyun luntan*), an overseas democracy-movement forum (*haiwai minyun luntan*), an oppose-corruption forum (*fanfubai luntan*), a CDP forum (*Zhongguo minzhudang luntan*), and an Internet "guerilla warfare team" forum (*wangluo youjidui luntan*). The Tibetan exile community also makes extensive use of BBS and chat rooms. Visitors can discuss issues

---

[50]Ibid.

[51]"Doctor Jailed for Promoting Falungong on Internet," http://www.june4.org/news/database/jan2000/doctorjailed.html

[52]Bobson Wong, "Chinese Democracy Activist Sentenced," Digital Freedom Network electronic newsletter, December 18, 2001.

[53]Michael Dorgan, "Critics of Taiwan Policy Outwit the Censor," *South China Morning Post*, May 25, 2000. One poster critical of PRC policy toward Taiwan wrote, "From the day Chen Shui-bian was elected, the mainland government's first reaction was 'wait and see.' Today, the period of waiting and seeing is over. Mainland leaders should understand that Chen Shui-bian refuses to be Chinese. The mainland leaders' behavior on the Taiwan issue has further made people understand who is the paper tiger."

[54]An index of the sites is available at http://dinfo.org/bbsindex.html.

of mutual concern on Tibet Online (www.tibet.org), TibetCentral (home.earthlink.net/~suevt/tibet.htm), TibetChat, TibetLink (www.tibetlink.com), and Worldbridges Tibet (www.Worldbridges.com/Tibet/). Because of consistent flooding attacks, Falungong no longer uses BBS or chat rooms, but instead uses e-mail lists.

BBS and chat rooms are particularly important media for communication of sensitive political views and coordination of protest activities among university students. Indeed, Chinese students have made extensive use of the Internet for these purposes several times in the past five years: during the 1996 Diaoyu Islands dispute; in the aftermath of the accidental U.S. bombing of the Chinese embassy in Belgrade in May 1999; following the murder in late May 2000 of Qiu Qingfeng, a Beijing University student; after the April 2001 collision of a U.S. EP-3 surveillance plane and a Chinese F-8 fighter; and following the September 11 terrorist attacks on the United States.[55]

The 1996 Diaoyu Islands Protests. During the summer of 1996, renewed friction related to the long-standing dispute between China and Japan over the Diaoyu Islands prompted an outpouring of nationalistic sentiment and unauthorized public protests in China. The official Chinese media shied away from discussion of these events, but students at several universities in Beijing used Internet bulletin board sites, chat rooms, and e-mail to disseminate information that was not carried in official media, to communicate their views about the issue, and to organize demonstrations. Senior Chinese leaders

---

[55]As illustrated by the 1996 Diaoyu Islands protests and the demonstrations that followed the accidental bombing of the Chinese embassy in Belgrade in 1999, nationalism is a double-edged sword for Chinese leaders. When Beijing has perceived promoting nationalist sentiment to be politically advantageous, the regime has taken the lead in attempting to stir up patriotic, and sometimes even xenophobic, feelings among the populace; but popular nationalism has in many cases placed the regime on the defensive. As one Chinese interlocutor told the authors, in its efforts to manipulate popular nationalism to its own advantage without losing control of the scope and direction of such potentially powerful forces, the Chinese government is "riding the tiger." Nationalism can be used to exert diplomatic leverage, to promote political and social cohesion, and to boost domestic support for the regime. At the same time, however, Chinese leaders recognize the difficulty of controlling, containing, and directing potent nationalistic feelings once they have been unleashed. Indeed, Beijing appears to be acutely aware of the risk that an upsurge of nationalist sentiment could turn against the regime, particularly if aroused citizens perceive Chinese leaders as irresolute in defending Chinese interests or insufficiently firm in responding to outside pressure.

reportedly became extremely alarmed after learning that a student at Beijing Aeronautics University had announced over the Internet a plan to use a remotely controlled airplane loaded with explosives to destroy the Japanese embassy in Beijing. Although official propaganda pronouncements in previous months had stoked anti-Japanese sentiment, Chinese leaders—worried that nationalistic outbursts could harm Sino-Japanese relations or even be directed against the regime for failing to more assertively press Chinese territorial claims—moved swiftly to discourage any further protests on the mainland.[56] Authorities temporarily shut down Internet bulletin board sites at several universities in Beijing and "advised" the organizer of the protest campaign to leave the capital for a brief "vacation" in remote Gansu Province.[57] In response to the incident, Vice Premier Li Lanqing, whose portfolio includes education issues, reportedly ordered universities to increase their control over student use of the Internet.[58]

The May 1999 Embassy Bombing Demonstrations. Many Chinese people, including nationalistic university students, were genuinely outraged in the aftermath of the May 1999 mistaken U.S. bombing of the Chinese embassy in Belgrade, which the official Chinese media portrayed as a deliberate act of aggression on the part of the United States. The Chinese leadership initially sought to encourage and facilitate student demonstrations both to send a political message to the United States and to stave off possible popular and intraelite criticism. With official support, large crowds protested outside the U.S. embassy in Beijing and at several U.S. consulates in China. After a few days, however, the demonstrations appeared to be on the verge of getting out of control, and with the tenth anniversary of the Tiananmen crackdown approaching, the authorities became increasingly concerned that protesters might shift the focus of their anger

---

[56]For an excellent study of nationalism and the Diaoyu Islands dispute, see Erica Strecker Downs and Phillip C. Saunders, "Legitimacy and the Limits of Nationalism: China and the Diaoyu Islands," *International Security*, Vol. 23, No. 3, Winter 1998/99, pp. 114–146.

[57]Chen Chiu, "University Students Transmit Messages on Defending the Diaoyu Islands Through the Internet, and the Authorities Are Shocked at This and Order the Strengthening of Control," *Sing Tao Jih Pao*, September 17, 1996, in FBIS, September 18, 1996.

[58]Ibid.

toward the Chinese government. Indeed, comments posted on some Internet sites—which assistants and advisors to Chinese leaders, or perhaps even some leaders themselves, may well have read—showed that while a great number of students were outraged at the United States,[59] quite a few were also angry with their own leaders.

Many of the Internet postings harshly criticized the United States, expressing "outrage" or "indignation."[60] On the web site of the official *Guangming Daily*, students from Wuhan University defiantly declared, "The Chinese people cannot be bullied, and the Chinese people cannot be insulted!!!!"[61] Some postings on other sites, however, excoriated Jiang Zemin and other senior leaders for failing to respond more forcefully to perceived U.S. bullying. Once again, after having fanned the flames of antiforeign sentiment through the propaganda apparatus, Beijing recognized that "riding the tiger" of nationalism carried serious foreign-policy and domestic political risks. In this case, the risks were probably far greater than those in the 1996 Diaoyu protests, largely because of the volatility of the issue, but in part also owing to the wider availability of the Internet. The authorities thus moved swiftly to halt the demonstrations and restore order on university campuses.

Protests over the May 2000 Murder of a Beijing University Student. In May 2000, Beijing University and Qinghua University students, among others, used the Internet to disseminate news about the murder of a female Beida (Beijing University) student, Qiu Qingfeng, who was killed on May 20 while returning to one of Beida's satellite campuses, and to organize mourning activities and demonstrations in response to the incident. Beijing University officials at first attempted to prevent information on the murder from becoming public, but on May 23, the news was posted on Beida's popular Triangle BBS. On

---

[59]For more on Chinese reactions to the embassy bombing, see Peter Hays Gries, "Tears of Rage: Chinese Nationalist Reactions to the Belgrade Embassy Bombing," *The China Journal*, No. 46, July 2001, pp. 25-43. The article draws upon hundreds of e-mails, letters, and faxes from students and other Chinese citizens that were posted on the web page of the official *Guangming Daily* during the embassy-bombing protests. Many of the writers charged that the bombing was intentional and that it was intended to bully or humiliate China.

[60]Ibid., p. 35.

[61]Ibid., p. 34.

that day, the number of users viewing the site skyrocketed from less than 700 the day before to nearly 12,000.[62] Within hours, news of the incident spread to BBS at other universities and to forums hosted by popular web sites such as Sohu.com. As Guobin Yang observes, students used bulletin board sites during the next several days to rapidly disseminate information on the murder, to formulate the demands they would present to school officials, to motivate fellow students and organize protest activities on several campuses, and to keep students and other Internet users across the country informed of events on their campuses in near-real time.[63] On May 29, a posting on the BBS where news of the murder first broke analyzed the impact of the Internet on the case, arguing that the Internet enabled students to "break through the deliberate control of information and suppression of memorial activities by the authorities. . . . [I]f it had not been for the Internet . . . this case would also have been covered up."[64] A memorial web site established in honor of the slain student had reportedly received over 24,000 hits by late May 2000.[65]

The April 2001 Airplane-Collision Incident. In April 2001, Chinese Internet users flocked to chat rooms and BBS to express their views on the collision of a U.S. surveillance plane and a Chinese fighter plane, which resulted in the death of the Chinese pilot and a diplomatic standoff between Beijing and Washington. In online postings, Internet users, including many who identified themselves as students, alternately vented their anger at the United States for causing the death of the Chinese pilot, Wang Wei,[66] and criticized their own leaders for what they perceived as a timid response in the face of an affront to China's national honor. As they had in the aftermath of the

---

[62]For an in-depth case study of student use of the Internet following the murder of Qiu Qingfeng, see Guobin Yang, "The Impact of the Internet on Civil Society in China: A Preliminary Assessment," pp. 35–42. Yang draws on interesting primary sources, including postings to BBS at several Chinese universities.

[63]Ibid., pp. 38–39.

[64]Ibid., p. 41. Yang notes that this posting from the Beijing University Triangle forum found its way onto Netease.com two days later, thus reaching many more Internet users throughout China.

[65]See http://cn.netor.com/index.asp

[66]Dozens of web sites memorializing Wang Wei quickly sprang up on the Internet. One of the most popular of these memorial sites, http://cn.netor.com/m/memorial.asp?BID=5661, has received more than 140,000 visitors.

bombing of China's embassy in Belgrade two years earlier, Chinese leaders once again faced a difficult balancing act in handling the April 2001 crisis. Through the official media, they encouraged a nationalistic response to perceived aggressive behavior on the part of the United States, while at the same time attempting to ensure that the regime itself would not become a target of the surging wave of indignation.

The September 11 Terrorist Attacks. In the aftermath of the September 11, 2001, terrorist attacks on the United States, Chinese Internet users, including university students from schools throughout China, once again flooded bulletin board sites and chat rooms with postings. Some of those posting messages expressed sympathy for the victims, but as was widely reported in Hong Kong and U.S. media, others gloated over an incident that in their view was the result of U.S. "hegemonism." This posed an unusual problem for Chinese decisionmakers. Accustomed to demands from U.S. NGOs and other free-speech advocates to relax its restrictions on the Internet, Beijing quickly realized that censoring anti-U.S. postings on the Internet would likely reduce negative publicity and improve its image in the United States. Given the volume of traffic on popular sites, it was undoubtedly no easy task to screen all of the postings and delete messages that contained offensive themes. The staff of *People's Daily Online* reported that in the week following the September 11 attacks on the United States, the Strong Country Forum was hosting as many as 25,000 visitors simultaneously during peak hours. Yet the site's webmasters managed to delete many postings that praised the terrorist attacks on New York and Washington, as well as a few that called on Beijing to attack the United States. They also blocked messages that were inconsistent with Chinese policy, including some that urged Chinese leaders to assist the United States in retaliating against Osama bin Laden and the Taliban.[67]

BBS and Chat Rooms as Barometers and Safety Valves. For officials in Beijing, bulletin board sites and chat rooms offer potential political advantages of two types. First, there is some evidence to indicate that government officials use popular sites such as the Strong Coun-

---

[67]"Chinese Organ Screens Web Site," *Far Eastern Economic Review*, September 21, 2001.

try Forum to gauge public opinion on a broad range of domestic and foreign-policy issues. When discussing the site with journalists last year, for example, the deputy director of *People's Daily Online* noted, "The government is interested in seeing people's views on events." Another example is provided by Harwit and Clark (2001), who relate a central government official's assertion that Premier Zhu Rongji adjusted his approach toward Japan in response to Internet postings that criticized him for being too accommodating during an October 2000 trip to Tokyo.[68] Second, at least some elements of the regime may also view bulletin board sites as an outlet for people wishing to express themselves. This is particularly significant in a country like China, where there are few institutionalized channels through which individuals can air their grievances. "People have opinions," said the deputy director of the *People's Daily* Internet edition. "People have a need to discuss ideas."[69] While the ultimate effects of two-way Internet communication in China remain to be seen, there have already been significant developments. Indeed, in a forthcoming study of the Internet and the development of civil society in China, Guobin Yang argues that through the use of means such as chat rooms and bulletin board sites, Chinese Internet users "are engaged in the discursive construction of an online public sphere." This may ultimately enable Chinese citizens to engage online and off in "a new type of political action, critical public debate."[70] Of course, it is not only two-way, interactive communication via the Internet that is important in this regard; one-way electronic communication is sometimes equally potent.

One-Way Communication

In addition to using two-way communication, dissidents, Falungong adherents, and Tibetan exiles also use several means of one-way Internet communication, including e-mail, web sites, and web-based magazines. Most of these are designed to provide information pas-

---

[68]Eric Harwit and Duncan Clark, "Shaping the Internet in China: Evolution of Political Control over Network Infrastructure and Content," p. 405.

[69]"China Tightens Grip on Booming Net Cafes," *South China Morning Post*, July 30, 2001.

[70]Guobin Yang, "The Impact of the Internet on Civil Society in China: A Preliminary Assessment," p. 24.

sively to those who actively seek it out. However, some new one-way strategies, such as e-mail spamming, enable groups to transmit uncensored information to an unprecedented number of people within China, and to provide recipients with "plausible deniability." In part because of dissident countermeasures (such as the use of different originating e-mail addresses for each message), the PRC is unable to stop these attempts to "break the information blockade."

One-Way Communication Via E-Mail. The first incident in which dissidents used mass e-mailing to send information to PRC Internet users occurred not long after the Internet became publicly available in China. On June 4, 1995, the sixth anniversary of the Tiananmen Square crackdown, an article by Chen Ziming was e-mailed to thousands of Chinese Internet users.[71] Then in late 1999, Wei Jingsheng sent an e-mail from Paris to five official e-mail addresses of the Beijing government to explain the meaning of freedom of expression protected by Article 19 of the Universal Declaration of Human Rights. He said in the e-mail that the Internet had become a useful tool for human-rights protectors.[72]

As the number of Internet users in China has grown and technology for mass e-mailing has progressed, spamming has become an increasingly potent weapon for overseas dissidents and free-speech advocates. The publishers of two Chinese-language electronic magazines, *Tunnel* (*Suidao*) and *VIP Reference* (*Da Cankao*), have mounted the best-publicized and most-sophisticated efforts. *Tunnel*, the first Chinese e-magazine aimed at a mainland audience, was launched on June 3, 1997. It is published weekly and is reportedly compiled and edited largely within China, then sent to Silicon Valley, and finally mass e-mailed back to the PRC from anonymous, U.S.-based e-mail accounts, such as nobody@usa.net. Sender names have included "Cyberspace Warrior" and "Digital Fighter."[73] According to the preface to the inaugural issue of *Tunnel*, the magazine's goal is to "break the current information blockade and suppression of speech

---

[71]Steve Usdin, "China Online," *Yahoo Internet Life*, January 1997.

[72]"Internet Used to Promote Freedom of Expression in China," Taiwan Central News Agency, October 2, 1999.

[73]William J. Dobson, "Dissidence in Cyberspace Worries Beijing," *San Jose Mercury News*, June 28, 1998; see also "China: Activists Launch Online Magazine," Reuters, June 18, 1997.

on the mainland." The editors opine that computers and the Internet have "enabled the technology for information dissemination to extend to everyone's desktop" and can thus be used to "disintegrate the two pillars of an autocratic society: monopoly and suppression."[74] The magazine contains articles on a wide variety of topics, from politics to economics to social issues, as well as occasional Chinese translations of Western press reports. Recent issues, for example, have carried articles on interest-group politics, tax revolts in rural China, corruption among Chinese government officials, and the business exploits of Jiang Mianheng, the son of Chinese President and Communist Party leader Jiang Zemin.[75]

The e-magazine *VIP Reference*, founded in November 1997, is edited and distributed by approximately one dozen overseas Chinese information-technology (IT) professionals, students, and academics based in Washington, D.C., and New York. It is published in a weekly edition, *Da Cankao*, and a daily edition, *Xiao Cankao*.[76] The editors describe themselves as Internet experts who support freedom of speech and declare that they are "destined to destroy the Chinese system of censorship over the Internet." *VIP Reference* contains articles from Hong Kong, Taiwan, and Western news sources that are not available to the public in China.[77] The magazine declines to reveal how many mainland subscribers it has, but it is reportedly sent to between 250,000 and 300,000 Chinese e-mail addresses. The number of recipients is likely to grow, as *VIP Reference* editor Richard Long

---

[74]*Suidao* [*Tunnel*], June 3, 1997, sd9706a. The first issue also contains an article on the Tiananmen crackdown by an author identified as "Temporarily Anonymous" (*Zan Wuming*).

[75]See, for example, *Suidao* [*Tunnel*], February 27, 2002, 0202a; *Suidao* [*Tunnel*], January 27, 2002, sd0201b; and *Suidao* [*Tunnel*], January 14, 2002, sd0201a.

[76]The name *VIP Reference* is apparently a play on the names of several classified compilations of translated Western news reports available only to senior party and government officials. These compilations include *Reference News* (*Cankao Xiaoxi*), and *Reference Materials* (*Cankao Ziliao*).

[77]"Frequently Asked Questions," version 2.0, March 14, 1998, available on the *Da Cankao* web site. The editors encourage recipients to redistribute *Da Cankao* but warn that they may be arrested for doing so, whereas they will not be arrested simply for receiving the magazine. The editors also note that they intend to begin publishing articles from underground writers in China.

(an alias, according to some reports) claims that he can now send 1,000,000 e-mail messages to China within a period of 10 hours.[78]

The editors of both *VIP Reference* and *Tunnel* employ a variety of countermeasures to protect readers and thwart any PRC efforts to prevent mainland users from accessing their publications. Both e-magazines attempt to provide a degree of "plausible deniability" to their subscribers by spamming tens of thousands of copies to recipients who have not requested them, including numerous CCP and Public Security Bureau officials. Even the head of the Shanghai Public Security Bureau's Internet security division reportedly receives a copy. The editors of *VIP Reference* also frequently change web site addresses and use different e-mail addresses every day to prevent Chinese security services from blocking distribution of their electronic publications. "This is like a war, [and] the Internet is the front line," says Lian Shengde, an information-systems specialist who works on the publication and distribution of *VIP Reference* at night.[79]

Following in the footsteps of *Tunnel* and *VIP Reference*, several other dissident organizations have attempted to mass e-mail information to mainland Internet users. New York–based members of the CDP, for example, have launched a campaign to send 100,000 copies of the banned party's political platform and other documents to mainland e-mail users.[80] CDP members in Britain, France, Germany, and Australia have also participated in the effort, which Xie Wanjun, director of the CDP's Internet department, describes as a form of "Internet guerilla warfare."[81]

---

[78]Kevin Platt, "China's 'Cybercops' Clamp Down." "The Internet is a revolutionary tool for people's freedom," Long proclaims. "China alone can't stop this global trend."

[79]Melinda Liu, "The Great Firewall of China," *Newsweek International*, October 11, 1999.

[80]The CDP Political Program (*Zhongguo minzhudang zhengzhi gangling*) calls for a variety of reforms, including land privatization (*tudi siyouhua*), freedom of the press and speech (*xinwen ziyou, yanlun ziyou*), constitutional democracy with separation of powers (*fenquan zhi de xianzheng minzhu tizhi*), judicial independence (*sifa duli*), placing the army under state rather than party control (*jundui guojiahua*), and allowing farmers and workers the right to organize independent associations to protect their rights and interests (*nongmin, gongren you quan zuzhi duli nonghui he duli gonghui baozhang qi quanyi*).

[81]"PRC Internet Police Said in On-Line Conflict with China Democracy Party," *Tai Yang Pao*, April 23, 2000, in FBIS, April 24, 2000. Chinese security services reportedly

Another type of one-way e-mail is employed by the International Campaign for Tibet (ICT): Visitors to the ICT web site can request to be added to the "Save Tibet E-mail Alert Network" and will there-upon receive a steady stream of e-mails from ICT containing recent statements or movements of the Dalai Lama, notices about the imprisonment and release of Tibetan political prisoners, and new Chinese efforts to repress Tibet. More important, the e-mails notify subscribers about current and future advocacy campaigns and provide information about how individuals can volunteer their time or financial contributions to the effort. Subscribers can even exercise a variant of the one-way e-mail effort by forwarding the e-mail alerts to their congressmen or other government officials.

One-Way Communication Via Web Sites. Dissidents, Falungong adherents, and Tibetan exiles also utilize web sites for communication and motivation. The overseas branch of the CDP, Frank Lu, and Falungong maintain particularly interesting and informative web sites (see the Appendix).

China Democracy Party. The CDP web page includes links to organizational information, important CDP documents, a publicity department, an invitation to join the CDP, and a variety of BBS forums (discussed above). The mainland and overseas organizational-structure (*guonei-wai zuzhi jigou*) area of the site lists information on more than 20 overseas departments, including an Internet department (*Zhongguo minzhudang haiwai wangluobu*) headed by Xie Wanjun, and the membership of the CDP Overseas Work Committee (*haiwai gonzuo weiyuanhui*). It also contains a list of 194 "open leading members" (*gongkai de lingdao chengyuan*) of CDP party branches in 29 provinces, municipalities, and special administrative regions, and one university on the mainland, as well as a listing of 64 CDP "secondary independent branches" in China. In addition, it includes a list of members of the CDP's national committee, with links to biographies of prominent members such as jailed CDP leaders Wang Youcai, Xu Wenli, and Qin Yongmin, as well as overseas committee members, including Xie Wanjun.

---

launched a technical counterattack against the CDP's mass e-mail campaign (discussed in more detail later in this chapter).

The core-documents section contains the complete Chinese text of the CDP's political program (*zhengzhi gangling*); the declaration of the party's founding in Zhejiang Province on June 25, 1998; an open letter to Jiang Zemin; a declaration of support for the U.S. granting of PNTR to China; and a statement on the eleventh anniversary of the Tiananmen crackdown written by a Shanghai CDP member.[82] The publicity department and "join the CDP" pages were both still under construction (*jianshe zhong*) when the authors last viewed the web site. The CDP home page provides a link to a password-protected "internal circular/secret code required" (*neibu tongbao-xu mima*) area.

Frank Siqing Lu, director and lone employee of the Hong Kong Information Center for Human Rights and Democracy, maintains a web page that features daily press releases on arrests of dissidents and practitioners of Falungong and Zhonggong, information on worker demonstrations, and links to news updates from a variety of international sources, such as the BBC and Radio Free Asia.[83] The site also contains Chinese- and English-language versions of an introduction to the center and its mission, along with a fundraising appeal, and invites readers to subscribe to *China Watch*, Lu's e-mail magazine.[84] In addition, there are links to longer reports on the Internet in China and on the suppression of religious and *qigong* organizations, as well as a special section on the crackdown against the Zhonggong *qigong* group. This special section contains a description of the group, a biography of its leader, and a database of news updates, which includes several documents that are identified as classified Public Security Bureau memos related to the crackdown on Zhonggong. (Lu's website, www.89-64.com, was hacked early this year. The site was replaced with a message that says, "This site saled to www.islam.org." In addition, a back-door trojan-horse virus is

---

[82]The CDP's "New Century Declaration," which was issued by the party's Beijing branch on December 31, 1999, is available elsewhere on the Internet in both Chinese and English.

[83]For a profile of Lu, see Maureen Pao, "Information Warrior," *Far Eastern Economic Review*, February 4, 1999, pp. 26–27.

[84]See www.89-64.com. Lu also brags on his web site that he is able to confirm and release news much more quickly than other activists or human-rights NGOs, and he claims that he is quoted by major media 10 times more frequently than the New York–based advocacy group HRIC.

automatically transmitted to the computers of Internet users who attempt to view Lu's website. The perpetrators have not been identified, but Lu is said to believe that the Public Security Bureau is responsible for the attack.)

Several other exile dissident organizations, including the CDJP and the Free China Movement, also maintain web sites.[85] Last year, mainland and exile dissidents jointly established a web site containing a database of information on corrupt Chinese officials, but the site has either moved or is no longer available.[86]

Falungong. Falungong has an extensive and highly organized network of global web sites. After Li Hongzhi arrived in the United States, he met with overseas Chinese who were followers of Falungong and knowledgeable about web-site design. After the first web site was posted, Falungong's online presence grew quite rapidly. The site (www. falundafa.org), which is bilingual, frequently updated, and well organized, contains messages from Li Hongzhi, a primer on the group's beliefs, links to 26 local Falungong web sites around the world, calendars of conferences and events, news items, and audio downloads that enable practitioners to listen to Master Li's lectures from anyplace in the world.[87] Li, who lives in Queens, reportedly makes his living mainly from Web-based sales of his book, *Zhuan Falun.*

Tibet. The Tibetan government-in-exile and its supporting NGOs maintain a sophisticated and informative set of web sites around the world (see the Appendix). The main advocacy sites can be divided into three general categories: official, supporters, and radicals. The main official site for the Dalai Lama's government (www.tibet.com) is maintained by the Office of Tibet in London. A companion site, www.tibetnews.com, is run by the Tibetan Government Department of Information and International Relations and serves as the government's official information site. Tibetnews publishes *Tibetan Bulletin,* the government's official online journal.

---

[85]See the Appendix for a comprehensive list of Chinese dissident web sites.

[86]"Non-Governmental Report Center Web Page Set Up," Hong Kong Information Center for Human Rights and Democracy, April 3, 1999.

[87]"Falungong and the Internet: A Marriage Made in Web Heaven," VirtualChina.com, July 30, 1999.

Among the hundreds of sites maintained by supporters of Tibetan independence, three illustrate the various uses of the web by the international NGO community.[88] The first is informational. Tibet Online (www.tibet.org), maintained collectively by the major Tibetan NGOs, is a comprehensive resource of information, Tibet support activities, and links. The site aims to "level the playing field by leveraging the Internet's ability to harness international grassroots support for Tibet's survival, while at the same time helping Tibetans involved in these efforts pick up highly valuable skills." The site offers English information on the global Tibetan support network, as well as links to 56 non-English sites in 18 languages. Other notable information sites include World Tibet News (www.tibet.ca), TibetLink (www.tibetlink.com), Current Tibet News from the International Campaign for Tibet, and the Tibet Information Network (www.tibetinfo.net).

A second supporter site seeks to motivate and coordinate Tibetan independence support activity. The International Campaign for Tibet (www.savetibet.org), located in Washington, D.C., is a primary focal point for coordination of various global campaigns. In June 2000, the ICT was running five major campaigns ("Stop PetroChina," "Free the Panchen Lama," "NO to Permanent NTR for China," "Election 2000," and a campaign directed at the World Bank), as well as championing the cause of 10 imprisoned activists and maintaining a constant congressional lobbying effort. Another notable motivational site (www.tibet.org/sft) is run by the Students for a Free Tibet, which has chapters in eight universities in the United States and Canada.

A third supporter site aims to raise money for the financing of pro-Tibetan activities. The self-described mission of the Tibet Fund (www.tibetfund.org) is "the preservation of the distinct cultural, religious, and national identity of the Tibetan people." Specifically, the Tibet Fund seeks to raise money, in the hopes of

> supporting economic and community development projects in the refugee communities in India and Nepal, providing support to monasteries and nunneries outside of Tibet, supporting projects to preserve Tibet's unique culture and arts, improving health condi-

---

[88] For a list of 375 functioning links to Tibet-related web sites, see http://www.geocities.com/Athens/Academy/9594/links.html.

tions in the refugee community, extending assistance for health, education, and small economic development projects inside Tibet, and offering scholarships and cultural exchange programs.

Tibet Fund's web site not only provides information for potential donors about the details of these efforts, but is equipped to permit online contributions.

In addition to official and supporter sites, the Tibetan community of web sites also includes more radical elements, in particular, groups that seek more democracy within the Tibetan government or that reject the Dalai Lama's calls for peaceful, nonviolent methods of resistance. Among the former, the Dharmsala-based Tibetan Center for Human Rights and Democracy web site (www.tchrd.org) contains newsletters, press releases, and documents. Specifically, the center publishes a regular biweekly English-language newsletter, entitled "Human Rights Update," which reports the latest human-rights violations in Tibet. It also produces documents and reports designed both to lobby international institutions, such as the United Nations, and to educate the Tibetan community-in-exile about the principles of human rights and democracy.

More-violent methods have been espoused by some leaders of the Tibetan Youth Congress (TYC), including armed attacks against Chinese military facilities in Tibet. These views are expressed by some subscribers to the TYC web site (www.tibetanyouthcongress.org), particularly in *Rangzen Magazine*, which is available online in both English and Tibetan. The reasons for these new attitudes are concisely explained by the president of the New Delhi TYC branch, Tenzin Phulchung: "Our struggle has been too passive. Many of our young people are losing patience."[89] According to one report, Phulchung reportedly views the Karmapa Lama, who now lives and grants audiences in a Dharmsala monastery, as "the wrathful manifestation of the Buddha," who could propel young Tibetans to take more direct action against China. Interviews with international Tibet activists in June 2000, however, discount the strength of this radical faction of the TYC and suggest that its proposed methods have been suppressed by a moderate majority.

---

[89]Rama Lakshmi, "Young Lama Inspires Tibetan Exile Youth," *Washington Post,* May 29, 2000.

One-Way Communication Via Web-Based Magazines. Exile dissident groups produce a variety of online Chinese-language magazines including *Beijing Spring* (*Beijing Zhichun*), *Democratic China* (*Minzhu Zhongguo*), and *New Century Net* (*Xin Shiji*) (see the Appendix for a more complete list). The monthly *New Century Net*, founded in June 1996 by former Shanghai *World Economic Herald* reporter Zhang Weiguo, is aimed explicitly at the Chinese mainland audience, and authors on the mainland have contributed articles via e-mail.[90] In addition, the China Development Union, a Beijing-based pro-reform intellectual group banned in October 1998, apparently published an online magazine about environmental issues, entitled *Consultations*. Finally, the Tibetan movement publishes quite a few online journals, including the *Tibetan Review* (English), *Tibetan Bulletin* (English, French, Hindi, and Chinese), *Rangzen Magazine* (English, Tibetan), and *Tibet Journal* (English).

MEASURING SUCCESS

Postings on Chinese-language BBS extol the potential of the Internet to disseminate dissenting political views, calling it "the modern big character poster" (*xiandai dazibao*) and proclaiming that "the most powerful weapon of the democratic revolution is information" (*minzhu geming zui youli de wuqi shi xinxi*). Indeed, many observers, and even some activists, have extremely high expectations about the potential of the Internet to degrade the CCP's control of information and ultimately to undermine its monopolization of political power. However, many dissidents and human-rights activists take a more conservative view, recognizing that while the Internet is a potential vehicle for communicating to millions of Chinese, it has not yet reached its full potential.[91] The present study concludes that political use of the Internet has further degraded the CCP's ability to control the flow of information into and within China but rejects hyperbolic claims that the arrival of the Internet in China will inexorably lead to the downfall of the CCP. It is not our intention to suggest that sudden, revolutionary change in China is the goal of the

---

[90]Zhang Weiguo, "Evading State Censorship," *China Rights Forum*, Human Rights in China, fall 1998. Zhang laments that he lacks the personnel and financial resources to e-mail *New Century Net* to subscribers in China.

[91]Interviews, U.S.-based Chinese activists, June 2000.

majority of Chinese dissidents or pro-democracy liberal intellectuals. Indeed, the most thoughtful activists argue that gradual pluralization is the preferred outcome.

It is still early in the game, and the prospects for using the Internet to encourage political pluralization and reform in China depend in part on the ability of dissidents to make effective and innovative use of the technology. The dissidents themselves recognize this challenge. One U.S.-based human-rights activist recently stated that making effective use of the Internet is the most important challenge facing the dissident movement.[92] In all, Internet communication is a tool that presents dissidents with a variety of new opportunities for disseminating information to a larger and more geographically dispersed audience than ever before, and potentially for organizing their activities in unprecedented ways. The latter possibility is perhaps the most important, but dissidents have yet to exploit its potential to the fullest. As one interviewee put it, the key to making effective political use of the Internet is finding ways to "turn information into action." Human-rights advocates cite the use of information technology by anti-Suharto demonstrators in Indonesia as an illustration of the Internet's potential in this regard.[93]

At the same time, enhanced communication does not always further the dissident cause; rather, in some cases, it serves as a potent new forum for discord and rivalry between various dissident factions. The Chinese dissident community, both on the mainland and in exile, has been plagued by disagreements over goals and tactics, disputes over the leadership of the movement, and myriad personal rivalries.[94] In early 1999, a *Far Eastern Economic Review* article identified 18 separate Chinese democracy organizations in exile, and stories are rife about the fissures and cleavages among these groups.[95] Some of this internecine warfare is no doubt motivated by competition for what is ultimately a limited amount of media attention, public noto-

---

[92]Ibid.

[93]Ibid.

[94]See for example Matt Forney, "Dissonant Dissent," *Far Eastern Economic Review,* June 5, 1997, pp. 28–32; and Erling Hoh, "Freedom's Factions," *Far Eastern Economic Review,* March 4, 1999, pp. 26–27.

[95]Hoh, "Freedom's Factions," p. 27.

riety, and financial support. Information technology may have increased the ease and rapidity of communication within the dissident community, but it has clearly not contributed to the resolution of these intracommunity disputes. On the contrary, the Internet has provided a new forum for dissidents to air grievances, level accusations, and launch *ad hominem* attacks against their rivals. Chinese-language BBS sites are filled with messages denouncing various dissidents, lamenting the ineffectuality—and sometimes questioning the patriotism—of the overseas democracy movement,[96] and making a variety of other charges and countercharges.[97] It is not possible to verify the identity of message posters on such sites, a problem that occasionally generates online debates about the true identity and motivation of posters. Some exile dissidents allege that Chinese intelligence agents posing as dissidents have posted messages on various sites to sow dissension within the ranks of the exile democracy movement. (For more on PRC efforts to use the Internet to exploit the dissident community's weaknesses, see the discussion of government counterstrategies below.)

FUTURE TRENDS

This study identifies five key future trends in dissident use of the Internet: First, in the short term, the Internet will require some human-rights NGOs and advocacy groups to change their traditional focus on reporting arrests. Second, it will permit small groups and individuals with limited resources to exert much greater influence than would otherwise be possible. Third, it appears likely that overseas dissidents, and perhaps even mainland dissidents, will engage in

---

[96]A recent message entitled "To Every Overseas Democratic Personage" illustrates the persistent divisions between the mainland and overseas dissident movements. The poster decries members of the exile dissident community as "Yankee's lackeys" and suggests that they engage in several alternative pastimes (none of which is suitable to mention here) rather than "interfering with Chinese affairs."

[97]The North America Free Talk Forum (*Beimei ziyou luntan*) is a frequent site of online dissident exile-community disputes. In late May 2001, for example, a message ostensibly posted by New York CDP member Xie Wanjun criticized veteran dissident Wang Xizhe for improperly carrying out activities in the name of the CDP without the approval of party authorities. The message accused Wang of "serving as a Taiwan dog" (*chongdang Taigou*) and "conspiring to damage and split the CDP" (*yinmou pohuai, fenlie Zhongguo minzhudang*). It must be noted, of course, that it is usually impossible to verify the identity of the individuals posting such messages.

more e-mail spamming campaigns in the near future. Fourth, dissidents may increasingly turn to emerging "peer-to-peer" technology to exchange information. Finally, dissidents and other unauthorized organizations will try to find new ways to exploit the Internet's motivational and organizational potential.

Because news about the arrests of dissidents can be spread so quickly on the Internet, U.S.-based human-rights advocacy groups, such as HRIC and Human Rights Watch Asia, are no longer primary sources for this type of information. Such groups are now essentially secondary sources. They can provide verification of information disseminated on the Internet or released by Frank Lu, and they can also produce in-depth reports and organize advocacy campaigns. As a result of this change, these groups are focusing more on providing quality content that will stand out amid the ever-growing welter of information available online and will thus have an impact in both the United States and China.[98]

Small groups of activists, and even individuals, can use the Internet as a "force multiplier" to exercise influence disproportionate to their limited manpower and financial resources. This is part of a broader trend, as exemplified by Nobel laureate Jody Williams, who successfully used the Internet to gain support for a treaty to ban landmines.[99] As one interviewee pointed out, however, the costs of designing, maintaining, and publicizing a web site represent a significant "barrier to entry" for many small groups and individual human-rights advocates interested in establishing a major on-line presence.[100]

The PRC will continue to be essentially powerless to prevent overseas activists from spamming uncensored news and political information to mainland e-mail users. For this reason, it is likely that a growing number of overseas activists will launch mass e-mail campaigns over the next several years. As one interviewee predicts, there may soon

---

[98]Interviews, U.S.-based Chinese dissidents and Western NGO activists, May–June 2000.

[99]Maxwell A. Cameron, Robert J. Lawson, and Brian W. Tomlin (eds.), *To Walk Without Fear: The Global Movement to Ban Landmines*, New York: Oxford University Press, 1999.

[100]Interview, exile dissident, June 2000.

be "hundreds of *Da Cankaos.*"[101] Indeed, it appears that there is already an emerging trend toward more groups and individuals becoming involved in this type of "Internet guerilla warfare." Frank Lu Siqing, director of the Hong Kong Information Center for Human Rights and Democracy,[102] is seeking funding for a plan to send human-rights information and daily news updates to 1,000,000 mainland e-mail users. According to one of our interviewees, an activist based in North America is also planning to send mass mailings to 1,000,000 PRC e-mail addresses.[103] In the future, spamming activities of this sort may not be limited to the exile dissident community. Activists on the mainland are reportedly compiling their own lists of e-mail addresses to use for future mass-mailing campaigns.[104]

Dissidents, Falungong practitioners, and other activists in the PRC and abroad may increasingly turn to emerging peer-to-peer technology to exchange information. (This topic is the subject of a separate, ongoing RAND study.) For example, information-sharing technologies like Gnutella and Freenet allow users to exchange files without using a central repository that could become a tempting target for the authorities to shut down.[105] Some of these technologies also have the potential to provide dissidents with a relatively high degree of anonymity. Discussions with pro-democracy activists and other

---

[101]Interview, exile dissident, June 2000.

[102]Lu's center was formerly known as the Hong Kong Center of Human Rights and Democratic Movement in China.

[103]Interview, Chinese human-rights activist based in North America, June 2000. Like the editors of *Tunnel* and *VIP Reference*, this activist intends to change ISPs and e-mail addresses frequently in order to prevent Chinese security services from interfering with his campaign.

[104]Jim Hu, "Ten Years Later: Chinese Dissidents Using Net," CNET News.com, June 8, 1999. It should be noted, of course, that there is no reason to believe that all of these plans will ultimately prove successful. The dissidents face determined opposition from the Chinese authorities. (The countermeasures Beijing employs are discussed in Chapter Two.)

[105]We are indebted to Bob Anderson for bringing this issue to our attention. For more information on Gnutella, see http://www.gnutella.com/ and http://www.gnutellanews.com/information/what_is_gnutella.shtml. For more information on Freenet, see www.freenetproject.org. The technical capabilities of Freenet are explained in Ian Clarke, Scott G. Miller, Theodore W. Wong, Oskar Sanderg, and Brandon Wiley, "Protecting Free Expression Online with Freenet," *IEEE Internet Computing*, January–February 2002, pp. 40–49. A Chinese version of Freenet has a Chinese-only web site at http://www.freenet-china.org.

proponents of freedom of speech on the Internet suggest that while some Chinese Internet users have downloaded documents such as the *Tiananmen Papers* by using peer-to-peer applications,[106] the use of the technology for the transmission of politically sensitive materials is not yet widespread.[107]

Finally, dissidents and civil-society groups will try to find new ways to exploit the organizational and motivational potential of the Internet. This is the critical challenge for those who seek to use the Internet to enhance their efforts to articulate controversial views on sensitive political, economic, and social subjects and to promote change. The Internet, according to Guobin Yang, has already "diversified and strengthened existing forms of social organization," enabled Chinese citizens to create "virtual communities," and "in ways previously unimaginable" linked Chinese civil-society associations, particularly environmental groups, with their global civil-society counterparts.[108] "Online protest," writes Yang, "represents the expansion of a contentious civil society in China."[109] But to realize the full potential of the Internet in this regard, dissidents and other social activists will have to devise inventive new ways to employ information technology and perhaps create organizational forms that are better suited to exploiting the opportunities presented by the Internet. One of the crucial mistakes of the CDP, for example, was the building of a hierarchical organization that did not permit its mem-

---

[106]It is difficult to determine how great a role peer-to-peer technology has played in the dissemination of the *Tiananmen Papers* over the Internet, but the complete text of the Chinese version of the book is available on Freenet. Moreover, in May 2001, *VIP Reference* published an article containing instructions on using Freenet to download copies of the *Tiananmen Papers* in Chinese. See *Da Cankao* [*VIP Reference*], No. 1196, May 5, 2001.

[107]It is important to note that the most popular uses of peer-to-peer technologies in China to date have not been political in nature. To the contrary, although some peer-to-peer applications, such as the Chinese version of Freenet, are designed specifically to combat censorship on the Internet and address privacy concerns, most Chinese Internet users are undoubtedly more interested in using peer-to-peer applications for entertainment purposes such as downloading MP3 music files. Nevertheless, some exiled dissidents, including the editors of *VIP Reference*, are looking to peer-to-peer technology as a potential tool for reaching Internet users in China.

[108]Guobin Yang, "The Impact of the Internet on Civil Society in China," p. 43.

[109]Ibid., p. 44.

bers to take full advantage of information technology and rendered the group vulnerable to Beijing's traditional strategy of counter-leadership targeting.

Chapter Two

GOVERNMENT COUNTERSTRATEGIES

BEIJING'S DILEMMA: CONTROL VERSUS MODERNIZATION

China faces a very modern paradox. The regime seems to believe that the Internet is a key engine of the New Economy, despite the burst of the Internet bubble and the dashed hopes of numerous Chinese "dotcom" companies, and that future economic growth in China will depend in large measure on the extent to which the country is integrated with the global information infrastructure. Economic growth is directly linked to social stability for the Beijing leadership, and absent communism or some other unifying ideology, maintenance of prosperity has become the linchpin of regime legitimacy and survival. Since economic growth has required modernization of China's relatively poor communications infrastructure, China has quickly become one of the world's largest consumers of information-related technologies. Moreover, Chinese leaders view the development of information technology, particularly the Internet, in China as an indispensable element of their quest for recognition as a great power. In the words of a recent *People's Daily* article, "The degree of development of information networking technology has become an important yardstick for measuring a country's modernization level and its comprehensive national strength."[1]

At the same time, however, China is still an authoritarian, single-party state with a regime whose continued rule relies on the sup-

---

[1]Commentator's article, "Using Legal Means to Guarantee and Promote Sound Development of Information Network," *People's Daily*, July 12, 2001, in FBIS, July 12, 2001.

pression of antiregime activities. The installation of an advanced telecommunications infrastructure to facilitate economic reform greatly complicates the state's pursuit of internal security.[2] The challenge for the regime, as Nina Hachigian puts it, is to "prevent this commercial goldmine from becoming political quicksand."[3]

Faced with these contradictory forces of openness and control, China has sought to strike a balance between the information-related needs of economic modernization and the security requirements of internal stability.

THE NATURE OF THE CHINESE INFORMATION SECURITY ENVIRONMENT

From public statements, policies, and actions, it is clear that the Chinese regime is anxious about the possible consequences of the country's information-technology modernization, in particular, the increasingly complex and challenging information security environment. The problem can be analyzed along two dimensions, one foreign and one domestic. The foreign dimension involves concerns about technology importation. Paranoia in China about "backdoored" foreign software and hardware is ubiquitous, bolstered by well-publicized cases involving user-ID tracking features of the Pentium III chip,[4] Windows 98,[5] and Windows 95,[6] as well as suspicions

---

[2] For a comprehensive rendition of this argument, see Andy Kennedy, "For China, the Tighter the Grip, the Weaker the Hand," *Washington Post*, January 17, 1999.

[3] Nina Hachigian, "China's Cyber-Strategy."

[4] Intel, the maker of the Pentium III chip, has admitted that each chip has a unique serial number that can be tracked as the user navigates the Internet. For a sample of Chinese analysis on the subject, see Cao Xueyi, "Here Comes the Wolf, Raise Your Hunting Rifle—Be Alert to Computer Network Security," *Jiefangjun bao*, August 25, 1999, p. 5, in FBIS, August 25, 1999.

[5] Microsoft has now verified that its Windows 98 operating system generates a unique identification number. As the user navigates the Internet, the operating system sends data to a Microsoft web site, where a database of user information is compiled. For a typical Chinese reaction, see Liu Youshui and Zhang Wusong, "High-Tech Development and State Security," *Jiefangjun bao*, January 11, 2000, p. 6, in FBIS, January 11, 2000.

[6] The Australian Navy allegedly discovered that their copies of Windows 95 were transmitting user information to Microsoft without the users' knowledge and subsequently accused Microsoft of attempting to steal naval secrets and undermine

that encryption products of U.S. origin have been deliberately weakened in a back-room export control deal with the U.S. National Security Agency.[7] Government officials, especially those in the security apparatus, seem convinced that foreign intelligence services are using or plan to use these "hidden dangers" (Chinese commentators also call them "time bombs") to the detriment of China.[8] According to one prominent Chinese information-security specialist:

> [D]uring the current phase of large-scale importation and use of foreign information equipment, we do not independently possess secure information systems, objectively speaking. For us, this is undoubtedly a great potential threat.[9]

This concern, coupled with the widespread belief that the U.S. military used back-doored hardware and software to advantage against Iraq and Serbia, has created the overall impression in Beijing that the importation of foreign equipment undermines Chinese national security and has led to calls for protectionist measures against foreign IT companies.[10]

The domestic dimension of the problem has been clearly articulated by the top leadership in China. Internal stability has been one of the state's top goals for the past few years and remains a crucial regime concern today.[11] Indeed, recent outbreaks of worker unrest sparked

---

Australian national security. For an understandably sympathetic Chinese spin on the incident, see Chen Ting and He Jing, "Pay Attention to Phenomenon of 'Information Colonialism,'" *Jiefangjun bao*, February 8, 2000.

[7]This assertion, a common theme in the Chinese press, is repeated in Xu Xiaofang and Dan Aidong, "Serious Challenge to Information Network Security," *Jiefangjun bao*, July 20, 1999, p. 6, in FBIS, July 20, 1999.

[8]The label "hidden dangers" is used in "Ensuring PRC Military Network Security," *Xiandai junshi*, October 11, 1999, pp. 35–36. The phrase "time bombs" is found in Xu Xiaofang and Dan Aidong, "Serious Challenge."

[9]Teng Yue, "China Should Handle Information Security Independently," *Wen wei po*, July 12, 1999, p. A7, in FBIS, July 27, 1999.

[10]Perhaps the most complete rendition of this argument, including analysis of alleged information warfare against Iraq and Serbia, can be found in Cao Xueyi, "Here Comes the Wolf."

[11]For articulations of this viewpoint over the past several years, see, for example, "Jiang Zemin Stresses Rural Stability at NPC Panel Discussions," *Xinhua*, March 5, 2002, in FBIS, March 5, 2002; "Zhu Rongji: Reform Requires Social Stability, State Security," *Xinhua*, March 5, 2000; "Jiang, Zhu Speak at Politics and Law Conference,"

by layoffs from failing state-owned enterprises (SOEs), and the expectation that such incidents are likely to become increasingly common in the wake of China's World Trade Organization (WTO) accession, have apparently intensified Beijing's concerns about social unrest and internal stability.[12] The government fears that hostile organizations, either foreign or indigenous, will use the new information technologies to agitate the population and undermine the regime. The following quote from the Public Security Bureau's official newspaper illustrates the regime's long-standing anxiety:

> As reform and the opening up [of] policy deepen, the problem of hostile elements at home and abroad using computer information systems and international networks to carry out infiltrations and sabotage will deteriorate.[13]

The cited possibilities are numerous. Chinese officials often warn that the information infrastructure could be used to disseminate information that is "harmful to social stability."[14] More alarming is the possibility that the new systems could be used to divulge state secrets or, worse still, to coordinate organized political action among antiregime forces.[15] Recent scandals involving the use of the Internet by the banned Falungong organization, the discovery of a high-ranking spy within the military,[16] the use of the Internet by antiregime dissidents, including the CDP, and a series of high-profile disclosures

---

Xinhua, December 23, 1998, in FBIS, December 24, 1998; and "Jiang Stresses Stability, Unity," *Xinhua*, December 18, 1998, in FBIS, December 18, 1998.

[12]For an excellent series of reports on the latest outbreaks of labor unrest, see John Pomfret, "With Carrots and Sticks, China Quiets Protesters," *Washington Post*, March 22, 2002; "China Cracks Down on Worker Protests," *Washington Post*, March 21, 2002; and "Thousands of Workers Protest in Chinese City," *Washington Post*, March 20, 2002.

[13]Sun Wen, "Use Computer to Fight Crime and Pornography," *Renmin gongan bao*, February 8, 1996, p. 3, in FBIS, February 8, 1996.

[14]Zhao Ying, "Information Security Issues," *Jingji guanli*, No.5, May 5, 1998, pp. 16–17.

[15]Willy Wo-Lap Lam, "Big Push to Maintain Stability in 1999," *South China Morning Post*, Jauary 5, 1999.

[16]Hua Chen, "PLA Spy Major General Liu Liankun Has Extensive Interpersonal Relationships and Strong Backing and His Execution Was Enforced by the Ministry of State Security," *Ming Pao*, September 19, 1999, p. A12, in FBIS, September 19, 1999.

of "state secrets" by the domestic media[17] have only served to heighten anxiety among government policymakers.

COUNTERSTRATEGIES

Beijing has employed two broad types of counterstrategies to deal with the potential threat of Internet use for dissident and other anti-regime activity. The first, which could be dubbed "low-tech solutions for high-tech problems," draws upon the state's Leninist roots and tried-and-true organizational methods. The second, "high-tech solutions for high-tech problems," embraces the new information technologies as an additional tool of state domination. The mixture of the two has proved to be a potent *yin* and *yang*, deterring most antiregime behavior and neutering whatever remains.

Low-Tech Solutions for High-Tech Problems

The low-tech solutions employed by the Chinese authorities include the use of informers and surveillance, arrests of Internet dissidents, promulgation of regulations, and, in some cases, the physical shut-down of network resources.

Informers and Surveillance. Most dissidents known to the authorities in China are subject to varying levels of intrusive surveillance, especially during sensitive political anniversaries or visits by sympathetic foreign dignitaries.[18] In late 1998, for example, the PRC authorities reportedly placed 150 dissidents on an intensive watch list. At least half of these dissidents, described as "dangerous" in internal directives, had affiliations with the banned CDP. The tightening of surveillance was allegedly prompted by the dissidents' ability to form an effective national network, which was particularly alarming to the authorities.[19]

---

[17]For one version of the Chinese response to these leaks, see "Jiang Zemin Orders State and Military Security Departments Guard Against the Leaking of State Secrets Via the Internet," *Ming Pao*, September 22, 1998, p. A14, in FBIS, September 22, 1998.

[18]See U.S. Department of State, *China Country Report on Human Rights Practices, 2000*, February 21, 2001, especially pp. 15–17.

[19]Willy Wo-Lap Lam, "Beijing Orders Close Watch on 150 Dissidents," *South China Morning Post*, December 24, 1998, p. 1.

Interviews suggest that this "nationalization" of previously localized movements was greatly aided by the Internet, which allowed dissidents to easily transcend physical borders and obstacles. Yet there is little evidence to suggest that Internet monitoring is a crucial part of the overall surveillance operation.[20] Instead, the authorities most often reportedly become aware of new dissidents or new activity by veteran dissidents through informers and other traditional means. Once a dissident or Falungong member has been identified as a person of particular interest, however, the authorities have the capability to monitor his or her electronic communications, including use of the Internet (see the discussion of e-mail monitoring and filtering below).[21]

Arrests and Seizures. The evidence suggests that once dissident activity has been identified, the security apparatus is more interested in obtaining physical than digital evidence. Searches and arrests are often conducted in the middle of the night to achieve maximum surprise. Western officials and human-rights monitors say that during searches of any political suspect's home or office, the first thing Chinese security agents seize these days is the computer, hoping to find on the hard drive incriminating evidence such as incoming or outgoing e-mail messages to co-conspirators.[22] Examples of this technique are numerous:

- In October 1998, Qin Yongmin was taken in for questioning, and police confiscated his computer and three fax machines.[23]

- On October 26, 1998, police detained the three leaders of the China Development Union (Peng Ming, Gan Quan, and Chang

---

[20]Some dissidents reportedly travel by train or bus to hold clandestine meetings with their colleagues in other cities, in part because they fear that the authorities routinely monitor their online communications. Ironically, the attempts of these activists to conceal their activities frequently play into the hands of the domestic and foreign intelligence services. Indeed, interviews indicate that communicating in person rather than electronically frequently leaves dissidents trapped in the Public Security Bureau's nationwide surveillance net.

[21]Interviews, PRC officials, Chinese dissidents, and Falungong practitioners, June–August 2000.

[22]Kevin Platt, "China's 'Cybercops' Clamp Down," p. 6.

[23]"Veteran Dissident Qin Yongmin Detained Again," *Agence France Presse*, October 27, 1998.

Qing), confiscating computers and documents from their head-quarters.[24]

- CDP leader Wang Youcai was detained on November 2, 1998, and formally charged on November 30, 1998. Wang was sentenced in Hangzhou Intermediate Court on December 21, 1998, to 11 years in jail for allegedly conspiring with foreign forces to overthrow the Chinese state. The specific charges against him included e-mailing 18 copies of the CDP constitution and declaration on the party's founding to dissidents and human-rights activists in the United States and Hong Kong[25] and accepting $800 to buy a computer.[26] Officers from the Hangzhou Public Security Bureau discovered the e-mail messages when they searched Wang's home and questioned him about the founding of the CDP in late June 1998.[27] The timing of these events supports the argument that PRC authorities rely primarily on traditional low-tech measures—in this case, a physical search of Wang's residence following his founding of the CDP—to uncover evidence of "subversive" Internet use. Specifically, the date the authorities claim to have uncovered Wang's e-mailing is the same as that of the raid on his apartment.

- In February 1999, Wang Ce, a prominent exiled democracy activist who clandestinely returned to China in 1998, was sentenced to four years in prison on charges of "abetting subversion" by giving CDP co-founder Wang Youcai $1,000 to purchase a computer.[28]

---

[24]Willy Wo-Lap Lam, "China Development Union Leaders Arrested, Told to Close," *South China Morning Post*, October 27, 1998, p. 9.

[25]"Hangzhou Court Verdict on Wang Youcai," Hong Kong Information Center, December 21, 1998, in FBIS, December 21, 1998. The court also noted that the Public Security Bureau found an e-mail message from a "hostile overseas organization" that had provided funding to Wang.

[26]Scott Savitt, "China's Internet Revolution," *Asian Wall Street Journal*, December 21, 1999, p. 10.

[27]According to the court verdict, "the public security organs testified that they found on the Internet and in 'Netscape Mail' of the defendant's Toshiba Satellite Pro 430 CDT on 26 June 1998 some 18 e-mail copies of the 'constitution' and 'declaration' sent by the defendant to overseas recipients on 25 June 1998."

[28]Authorities also charged Wang Ce with "illegally entering the country."

- In February 1999, Xu Wenli's wife, He Xintong, demanded the return of some unlisted items among the 300 items seized from their home during three police searches. The missing items included two computers, two printers, two telex machines, a photocopier, and numerous cassettes and CDs.[29]

- On June 19, 1999, 15 police stormed into Zhu Yufu's home and hauled him and another CDP member, Han Shen, away. The police searched his home for two hours, taking away a computer, an address book, and numerous documents.[30]

- On June 29, 1999, Gao Hongmin, deputy chairman of the Beijing-Tianjin branch of the CDP, was taken from his home by police, who also removed his computer and documents.[31]

- In August 1999, the wife of Wu Yilong (Shan Chenfeng) reportedly was detained. The authorities confiscated a computer, an address book, a fax machine, books, and other items.[32]

- In September 1999, Qi Yanchen, an employee of the Hebei branch of the China Agricultural Development Bank and a member of the China Development Union (CDU), a banned pro-reform intellectual group,[33] was arrested for his involvement in a variety of "subversive" online activities. These reportedly included posting portions of his unpublished book, "China's Collapse," on overseas Chinese-language BBS. The book explored themes including social instability in China and warned that the ruling CCP needed to enact political reform or risk turmoil.[34] He was also accused of publishing articles under the pseudonym "Ji

---

[29]"Wife Demands Return of Confiscated Items," *Agence France Presse*, February 18, 1999.

[30]"Hangzhou Security Bureau Detains Five More Dissidents," Hong Kong Information Center of Human Rights and Democratic Movement in China, June 19, 1999.

[31]"Two Democracy Party Members Detained," *Agence France Presse*, June 29, 1999.

[32]U.S. Department of State, *Human Rights Report for 1999*.

[33]Peng Ming founded the CDU in early 1998. Beijing declared the organization illegal in October 1998 and subsequently sentenced Peng to 18 months of reeducation through labor on charges of soliciting a prostitute.

[34]"China Charges Dissident Author with Subversion," Associated Press, December 22, 1999.

Li" in *VIP Reference* and of receiving copies of *VIP Reference*.[35] According to other reports, his arrest may also have been related to his involvement with *Consultations*, an environmentalist on-line magazine associated with the CDU. Qi had been under police surveillance since 1998. At the time of his arrest, his computer, fax machine, printer, books, manuscripts, and notes, as well as copies of *VIP Reference*, were allegedly confiscated.[36]

- In November 1999, Zhejiang Province CDP members Wu Yilong, Mao Qingxiang, Zhu Yufu, and Xu Guang received sentences of 11, 8, 7, and 5 years, respectively, on charges that included using e-mail to communicate with "reactionary organizations abroad" and posting CDP materials on overseas Chinese-language BBS.[37] Authorities confiscated a computer belonging to one of the four dissidents when they detained them earlier in the year.

- In March 2001, state security officers in Beijing detained Yang Zili, a software engineer and outspoken critic of the CCP who maintained a website called "Yang Zili's Garden of Ideas." They also detained his wife, Lu Kun. The agents confiscated Yang's computer, books, and other items. Lu was subsequently released, but Yang reportedly remains in detention.[38]

Finally, the authorities have a clear track record of arresting possible dissidents for Internet-related offenses. The first person imprisoned in the PRC for "subversive" use of the Internet was Lin Hai, a computer software engineer and Internet entrepreneur from Shanghai, who was charged with subversion and was sentenced to two years in prison on January 20, 1999, for providing a total of 30,000 e-mail addresses to "overseas hostile publications." Authorities charged that Lin, using the on-line pseudonym "Black Eyes," began transmitting

---

[35]"Chinese Internet Writer Faces Trial for Subverting State Power," e-mail press release from *VIP Reference* editor Richard Long, May 22, 2000.

[36]U.S. Department of State, *Human Rights Report for 1999*. See also "Chinese Intellectual Detained for Alleged Internet Crimes," *Inside China Today*, www.insidechina.com/news, September 6, 1999.

[37]"Four CDP Founders Given Stiff Prison Sentences," Hong Kong Information Center for Human Rights and Democracy, November 9, 1999, in FBIS, November 9, 1999.

[38]For a detailed account of the detention, see Lu Kun, "My Experience in a Beijing Detention Center," April 13, 2001, available on the Digital Freedom Network website at http://www.dfn.org/voices/china/lukun-detention.htm.

the e-mail addresses of Internet users in Chengdu, Guangzhou, Shenzhen, Zhuhai, Zhanjiang, Huizhou, Shantou, Qingdao, and Shanghai to *VIP Reference* and *Tunnel* in September 1997. In January 1998, Lin sent more e-mail addresses to the editors of *VIP Reference* in response to their specific requests for the addresses of Internet users in Nanjing and Beijing.[39]

Lin argued that he transmitted the e-mail addresses solely for commercial reasons, but the court rejected his defense on the basis of his e-mail correspondence with *Tunnel* and *VIP Reference*, which suggested that he had political motivations.[40] Lin believes that the Public Security Bureau found out he had sent the e-mail addresses to the online magazines through e-mail filtering and then traced the e-mail address he used, which was hosted by a Chinese ISP.[41] However, available evidence is not sufficient to determine conclusively whether authorities initially discovered Lin's activities through technical monitoring or through more traditional means, such as the use of informants.[42]

Many observers said that Lin's sentencing reflected Beijing's growing unease about the potential of the Internet and e-mail to aid dissidents' efforts to organize, contact "overseas hostile forces," and disseminate uncensored information within China. The sentence was apparently intended to deter other would-be "cyber-dissidents" from

---

[39]"Court Verdict on Dissident Lin Hai," Hong Kong Information Center, January 20, 1999. Police originally detained Lin in March 1998. The charges brought against him also included forwarding copies of *Da Cankao* to a former classmate in Beijing and telling him how to subscribe to the magazine via e-mail.

[40]According to the authorities, in one message, Lin wrote, "Your electronic periodical is indeed a high-class magazine carrying an independent voice. Although I hate the tweeters of the CPC, I am alone and there is nothing I can do."

[41]Interviews, March 2002.

[42]Evidence presented by the prosecution at Lin's trial included two PCs, one laptop, and one modem; e-mail correspondence between Lin and editors of *Tunnel* and *VIP Reference*; e-mail correspondence between Lin and a former schoolmate; "relevant reports" from "data communication bureaus" in Guangzhou, Shenzhen, Zhuhai, Shantou, Zhanjiang, and Huizhou, the Shanghai Telecommunications Bureau, the Beijing Zhongxi Electronic Engineering Technology Developing Company, the Sichuan Public Information Industry Company, Limited, and the Shandong Qingdao Information Industry Company, Limited; an "Internet User Application Card" from the Shanghai Telecommunications Bureau; other unspecified written evidence; and "witnesses' statements."

using the Internet for subversive purposes. For unknown reasons, Lin was released early, in September 1999.[43] Despite his ordeal, Lin subsequently told the Associated Press that he was looking for new Internet-related business opportunities because "this business is very hot at the moment," but he allowed that "whether or not I can continue in this line of work depends on the political environment."[44] In 2000, Lin apparently reopened his web site, which advertised a list of more than 1,000,000 mainland e-mail addresses as well as a variety of information-technology services.[45] He ultimately left China and is now living in the United States.

In June 2000, police in Chengdu, Sichuan, arrested and charged with subversion the operator of a mainland web site that posted news about dissidents and the 1989 Tiananmen massacre. Huang Qi, who operated the www.6-4tianwang.com site, was seized by police on June 3, 2000, and held at a detention center in Chengdu. His wife, Zeng Li, who was taken away the same day, was released. Minutes before he was arrested, Huang posted messages on the chat room of his web site, saying that four policemen had come to take him away.[46] The web site was originally launched in June 1999 as the first in China dedicated to helping people find relatives abducted by traffickers. It was shut down in March 2000 over reports concerning the human rights of Chinese laborers working overseas. It was reopened in April 2000 with the help of a U.S.-based Chinese group.[47]

Many more Chinese dissidents, Falungong practitioners, and others have been charged with crimes related to political use of the Internet. Over the past two years, at least 25 people have been detained in

---

[43]See "PRC's Cyber-Dissident Released from Jail Early," Hong Kong *Agence France Presse*, March 3, 2000; and "China Grants Early Release to Cyberdissident," Associated Press, March 3, 2000. Lin's early release was not widely reported in Western media until March 2000. He was apparently reluctant to discuss with reporters the circumstances surrounding his early release, telling *Agence France Presse* only that "the real reason, nobody knows."

[44]"China Grants Early Release to Cyberdissident," Associated Press, March 3, 2000.

[45]The site, home4u.china.com/technology/internet/hopy/, is no longer accessible.

[46]Josephine Ma, "Police Charge Web Site's Founder with Subversion," *South China Morning Post*, June 8, 2000.

[47]Josephine Ma, "Defiant Cyber-Surfers Play Cat-and-Mouse Game," *South China Morning Post*, June 8, 2000.

China for their use of the Internet, according to the Digital Freedom Network.[48] Representative cases include the following:

- Six Falungong practitioners, four of whom were graduate students in engineering and sciences at Beijing's elite Qinghua University, were sentenced in December 2001 to prison terms of from three to 12 years for disseminating Falungong materials on the Internet.[49]

- That same month, Wang Jinbo, a member of the CDP in Shandong Province, was convicted of subversion and sentenced to four years in prison for e-mailing articles calling for the reversal of the official verdict on the 1989 Tiananmen democracy movement to overseas dissident groups.

- In September 2001, a Hunan Internet user was sentenced to three years in jail for e-mailing articles critical of the government to friends.

- A Chinese reporter affiliated with the CDP was sentenced in August 2001 to reeducation through labor for trying to recruit new members for the banned opposition party in Shanghai and posting pro-democracy articles on web sites.

- Li Hongmin of Hunan was detained in June 2001 for e-mailing excerpts from the Chinese version of the *Tiananmen Papers*, which was swiftly banned in China, to several friends.

---

[48]Digital Freedom Network, "Attacks on the Internet in China: Chinese Individuals Currently Detained for Online Political or Religious Activity," available on the DFN website at http://www.dfn.org/focus/china/netattack.htm. See also Digital Freedom Network, "Attacks on the Internet in China: Internet-Related Legal Actions and Site Shutdowns Since January 2000," available at http://www.dfn.org/focus/china/shutdown.htm.

[49]"Six Falungong Academics Jailed," *South China Morning Post,* December 24, 2001. The Falungong members had also distributed printed materials on the streets in Beijing, and it is not clear how the authorities initially discovered their activities. If this case followed the usual pattern, the police may have arrested the Falungong members after receiving a tip from an informant or learning that they were distributing the leaflets. The authorities probably confirmed that they were using the Internet to transmit similar materials only after arresting them and confiscating their computers. The possibility that the police observed and traced their online activities, however, cannot be discounted on the basis of the evidence available. In all of the incidents listed here, as in this case, it is not known how the authorities discovered the Internet activities of the individuals who were jailed or detained.

- In October 1999, Zhang Ji, 20, a student at Qiqihar University in Heilongjiang, was arrested and charged with "disseminating reactionary documents via the Internet." Zhang had reportedly transmitted news of the crackdown to Falungong members in the United States and Canada.[50]

It should also be noted that the authorities appear willing to charge dissidents with "subversive" uses of the Internet that are inherently nonpolitical in nature, primarily as a tactic to silence them or smear their character. In January 2000, Public Security Bureau officers arrested Wang Yiliang, a dissident writer from Shanghai, for his participation in an unauthorized literary association. The authorities searched his home and found pictures of nude women downloaded from the Internet on his computer, which they subsequently used to sentence him to two years of reeducation through labor for "possessing pornographic articles."[51]

Promulgation of Regulations. One of the most effective lines of defense in China's Internet security strategy is the use of bureaucratic regulations to shape the market environment and the incentives of key participants in ways favorable to the state's interest. Since 1995, the Chinese government has promulgated a blizzard of rules covering nearly every aspect of the Internet market. In particular, the 1997 Public Security Bureau regulation entitled "Computer Information Network and Internet Security, Protection and Management Regulations" places most of the onus for monitoring, reporting, and preventing antiregime use of the Internet on domestic providers. At the level of international gateways, units that oversee international connections of networks "must assume responsibility for the Internet network gateways as well as the security, protection, and management of the subordinate networks" (Article 10), including technical security measures, education and training, and user registration. Below the gateway level, networks should, within 30 days of the opening of a network connection, carry out the proper registration procedures with a unit designated by the Public Security Bureau of the province, autonomous region, or municipality directly under the

---

[50]"China Charges Student on Falungong E-mail," Reuters, November 8, 1999.

[51]"Guizhou Poet Ma Zhe Has Been Sentenced to Five Years' Imprisonment on Subversion Charge," Hong Kong Information Center, March 14, 2000.

Central Government people's government. They are also required to give information to their local Public Security Bureau on the units and individuals that have connections to the network, as well as to keep the Public Security Bureau informed of any changes in the information about units or individuals using the network. Interviews in Beijing also confirm that the Public Security Bureau frequently requests information on customer databases from ISPs.[52] If a violation occurs, the ISP is required to assist the Public Security Bureau in investigating the incident and criminal activities involving computer information networks. Like network security, responsibility for maintaining database security resides with the ISPs, and violations by users result in cancellation of the ISP's business license and its network registration (Articles 20–23). As a result, ISPs have implemented certain self-censoring policies to avoid the wrath of the authorities. According to a January 24, 2000, Reuters story, ISPs pay employees known as "Big Mamas" to lead "armies" of volunteers who patrol chat rooms and bulletin boards, ferreting out risky political commentary, foul language, and unwanted advertisements.

The 1997 regulations also circumscribe the acceptable uses of the Internet by the users themselves. Users are required to register with the Public Security Bureau, filling out an application form that links their personal information with their network account information (it is important to note, though, that this rule is almost universally ignored). Once online, users are not permitted to use the Internet to create, replicate, retrieve, or transmit information with the following goals:

1. Inciting to resist or breaking the Constitution or laws or the implementation of administrative regulations.

2. Inciting to overthrow the government or the socialist system.

3. Inciting division of the country, harming national unification.

4. Inciting hatred or discrimination among nationalities or harming the unity of the nationalities.

5. Making falsehoods or distorting the truth, or spreading rumors destroying the order of society.

---

[52]Interviews, Chinese and Western executives, January 2000.

6. Promoting feudal superstitions, sexually suggestive material, gambling, violence, or murder.

7. Inciting terrorism or inciting others to criminal activity; openly insulting other people or distorting the truth to slander people.

8. Injuring the reputation of state organs.

9. Engaging in other activities against the Constitution, laws, or administrative regulations.

Moreover, the rules also bar anyone from using computer networks to "harm national security, disclose state secrets, harm the interests of the State, of society or of a group, the legal rights of citizens, or to take part in criminal activities" (Article 4).

A more recent regulation (State Council Order #273, "Regulations on the Management of Commercial-Use Encryption") attempts to supplement the existing Internet rules by strictly regulating the importation of encryption products, which the authorities fear might be used by dissidents, foreign spies, and other insurgent elements to hide their Internet-based activities from the prying eyes of the authorities. The authorities ultimately backed down from this regulation. Ironically, the fears of the security apparatus about encrypted e-mail currently appear unfounded, although this situation could change as e-commerce and other encryption-heavy applications become more widespread. Interviews suggest that few, if any, people in China—either citizens or foreigners—are using encryption to conceal the content of their communications, and this is especially true for dissidents.[53] In the words of one interviewee, "Why would I use encryption? Why would I want to call attention to myself?" Indeed, the reluctance to use encryption appears to dovetail with the widespread perception that the government is monitoring e-mail and other electronic communications for keywords, despite the seemingly insurmountable technical challenges of searching so much traffic without bringing the network to a grinding halt.

---

[53]Interviews, Chinese and Western sources, January 2000. Similarly, none of our contacts ever mentioned dissident use of steganography, a technique that can be used to conceal information by hiding text messages in images and audio or video files.

The Ministry of Public Security also allegedly sought to restrict the use of the Internet by dissidents prior to June 4, 2000. In June, the Ministry allegedly issued Document #33, calling on the Computer Management and Supervision Departments of the Public Security Bureaus in various localities to strictly manage Internet users and restrict them from browsing "reactionary" information. After receiving this circular, Public Security Bureaus in some provinces and cities held emergency meetings to coordinate control measures. They called on ISPs to place more overseas web sites on the list of blocked sites and demanded that web cafes be placed under strict control. For example, the Ministry of Public Security authorities in Baoding City, Hebei Province, issued new regulations for 100 Internet cafes in the city. The regulations announced a "point system," whereby a cafe "allowing a customer to browse 'reactionary' information will be deducted 10 points." If a cafe loses 30 points within a single year, its license will be suspended for one year.[54]

The authorities have continued to issue rules and regulations in response to perceived Internet-enabled challenges to the regime. In November 2000, the Ministry of Information Industry issued regulations governing the management of Internet bulletin board sites and chat rooms. The regulations mandated that operators of any of these types of "electronic public notice services" must register with the authorities, obtain a permit, enact measures to control content, and maintain records of all postings for 60 days. In addition, the regulations forbid Internet users from posting on such sites any information that endangers state security, harms national reunification, runs counter to the provisions of the PRC Constitution, exposes state secrets, undermines national unity, spreads heretical beliefs, disrupts social order, or is otherwise banned. Moreover, the regulations stipulate that ISPs must delete postings that fall into any of these forbidden categories and must report them to the authorities.[55]

---

[54]"Eight Reactionary Activists from Jilin Who Were Imprisoned Before Call for Reversing the Verdict on the June 4th Incident and for Government Compensation," Hong Kong Information Center for Human Rights and Democracy, June 3, 2000, in FBIS, June 3, 2000.

[55]"China Enforces Control of Electronic Public Notice Service over the Internet," *Xinhua*, November 6, 2000.

Several bulletin board sites have been closed for violating the regulations. In September 2001, for example, authorities temporarily shut down a popular BBS hosted by the Huazhong University of Science and Technology in Wuhan, after students posted articles about the 1989 Tiananmen Square crackdown. "Everybody knows universities aren't quiet places, but it's still too sensitive for students to talk about this on the Web," a local official told Western journalists. "Articles on the June 4 incident are a serious breach of state regulations on Internet information," he said.[56] In addition to issuing rules concerning BBS, Beijing also promulgated regulations in November 2000 that require popular sites like Sohu.com and Sina.com to carry only news content that has been approved by state-run media organs.

The most recent regulations on the use of the Internet were promulgated by the Ministry of Information Industry in January 2002. These regulations require ISPs to maintain detailed records about their users, install software to record e-mail messages sent and received by their users, and send copies of any e-mails that violate PRC law to the appropriate Chinese government departments.[57] Like many of the previously issued regulations, the January 2002 regulations place the burden of policing Internet use squarely on the shoulders of Internet industry companies.

In March 2002, the official *People's Daily* reported that more than 100 Chinese Internet industry executives signed a voluntary "self-discipline" pact aimed at promoting "the healthy and orderly development" of the Internet in China through "protection of intellectual property, network security and the elimination of deleterious information from the Internet."[58] The agreement pledges the com-

---

[56]"Student Net Site Closed over Talk of Tiananmen," Reuters, September 6, 2001. The *Baiyun huanghe* bulletin board site (bbs.whnet.edu.cn) was reopened after the university's Communist Party Committee expunged the messages about the Tiananmen crackdown, replaced student webmasters, and required all users posting messages to register under their real names. The authors last accessed the site in January 2002.

[57]For more on the latest regulations, see "China Sets New Net Rules," *South China Morning Post*, January 21, 2002.

[58]For the English version, see "China's Internet Industry Wants Self-Discipline," *People's Daily Online*, March 26, 2002, available at http://english.peopledaily.com.cn/200203/26/print20020326_92885.html; for the Chinese version, see Chen Jian, "*Zhongguo hulianwang xingye zilu gongyue zhengshi qianshu*" [Chinese Internet

panies of China's Internet industry to abide by all laws, regulations, and policies governing the development and management of the Internet, including those concerning online news. The signatories also promise to prevent the online publication or transmission of information that is "harmful to state security or social stability, violates laws and regulations," or is deemed "superstitious or obscene." In addition, they pledge to scrutinize the information users publish online and quickly remove any "harmful information" posted on their web sites that could have a "negative influence" on Chinese Internet users. The agreement also stipulates that portals will not link to web sites that contain "harmful information."[59]

When regulations and the threat of arrest are not enough, Beijing has shown that, at least under what it perceives to be pressing circumstances, it is willing to resort to more disruptive measures.

Physical Shutdown of Network Resources. In serious crises, the Chinese government has consistently been willing to shut down networks temporarily in order to gain control. The first harbinger of this trend occurred during the anti-Japanese demonstrations of 1996 that were sparked by perceived violations of the territorial sovereignty of the contested Diaoyutai/Senkaku Islands. When Chinese students used e-mail to organize anti-Japanese demonstrations, Chinese officials responded by shutting down computer bulletin boards on some campuses.[60] More recently, after the April 1999 Falungong gathering in Beijing, the government reportedly ordered one ISP to suspend e-mail service for two days. In the run-up to June 4, 2000, the Ministry of Public Security reportedly sought to slow down or shut down China's free and anonymous e-mail sites, 163.net and 263.net. The

Industry Self-Discipline Pact Formally Signed], *Renminwang*, March 26, 2002, available at http://www.people.com.cn/GB/it/49/149/20020326/695393.html.

[59]For the full text of the agreement in Chinese, see *"Zhongguo hulianwang xingye zilu gongyue"* [Chinese Internet Industry Self-Discipline Pact], *Renminwang*, March 27, 2002, available at http://www.people.com.cn/GB/it/49/149/20020327/695927.html. The pact also includes provisions on competition in the Internet industry and Internet user privacy issues, among other matters. Interestingly, it also contains a provision that says the signatories will propose recommendations for legislation and policy initiatives related to the development and management of the Chinese Internet industry.

[60]Steve Mufson, "Chinese Protest Finds a Path on the Internet," *Washington Post*, September 17, 1996, p. A9.

owners of 163.net, Liang Liwei and Leng Wanbao, said that they could neither send nor receive e-mail beginning on June 1. The government's reasoning for the repeated use of this tactic is neatly summarized by Pan Weimin, an electrical engineering graduate from Fudan and the head of operations for PaCity Computer Company: "When push comes to shove, the authorities don't have to restrict themselves to imposing a NetWall around China. They can use tried and true traditional methods: one administrative order from on high and everything can be shut down. It's simple and effective."

High-Tech Solutions for High-Tech Problems

In addition to traditional methods of control, the Chinese authorities have also made use of high-tech countermeasures, such as blocking websites and e-mail, government-sponsored hacking, monitoring and filtering of e-mail, and online propaganda, denial, deception, and disinformation.

Blocking Web Sites and E-mail. One of the most common, and perhaps most quixotic, methods employed by Beijing to stem the flow of antiregime information into China consists of the blocking of web sites and e-mail. Authorities at various times have blocked politically "sensitive" web sites, including those of dissident groups and major foreign news organizations, such as the Voice of America, the *Washington Post,* the *New York Times,* and the BBC.[61] In early 1999, Hong Kong–based activist Lau San Ching helped establish a web site commemorating the victims of Tiananmen (www.june4.org), and Beijing quickly blocked access to the site.[62] In April of the same year, stunned party bosses responded to the gathering of thousands of Falungong members in central Beijing by ordering the arrest and

---

[61]The *New York Times* web site was unblocked in 2001 after its reporters raised the issue with Jiang Zemin during an interview. The site remained accessible from China as of the completion of this study in January 2002. In addition, we note that the blocking is rather inconsistent—materials blocked on one web site often remain accessible on other sites. For example, the *Foreign Affairs* web site was blocked in early 2001, presumably to prevent Chinese Internet users from reading an article on the *Tiananmen Papers* that appeared there. But at the same time, the offending article was still readily accessible via a link from the Council on Foreign Relations web site, which was not blocked.

[62]Liu, "The Great Firewall of China."

prosecution of Falungong leaders and members and the blocking of access to the group's international constellation of web sites.[63] The authorities even dispatched censors to screen all Chinese web forums and bulletin boards and erase any favorable remarks about Falungong founder Li Hongzhi, whom they denounced as a charlatan and a doomsayer.[64]

Most of the government's attempts to prevent the viewing of banned web sites use multiple layers of filtering, ranging from the ISP to the network carrier to the nontechnical aspects of web surfing in China (e.g., registration with the police, observation by Internet cafe workers). The technical blocks themselves are carried out at the ISP or international-gateway level and involve the alteration of network routing tables. According to interviews, the four major ISPs regularly exchange information about which sites they are blocking. At the network-carrier level, CERNET apparently uses only CISCO routers with ACR (access-control routers), and site restrictions are set by the Ministry of Education. Most national-level blocks are placed by China's International Connection Bureau (ICB), a set of computers belonging to state-owned China Telecom. Software at that level is programmed to reject requests for access to banned sites. China blocks sites only if a large enough number of users access them. Blocking can be done only intermittently because the software does not have enough computing power to block every objectionable site all the time. By blocking a site intermittently, the government hopes users will simply assume that the site isn't accessible.[65] In addition, blocks on web sites are often temporarily removed when high-level foreign delegations visit China. During the October 2001 Asia-Pacific Economic Cooperation (APEC) meeting in Shanghai, for instance, Beijing permitted access to several web sites that are normally blocked, including those of the *Washington Post* and CNN. This was presumably intended to avoid embarrassing international media reports and to burnish China's image in order to make a good impression on visiting world leaders. As soon as President Bush and

---

[63]Ibid.

[64]Ibid.

[65]Julie Schmit and Paul Wiseman, "Surfing the Dragon: Web Surfers Find Cracks in Wall of Official China," *USA Today*, March 15, 2000, p. 01B.

other world leaders left Shanghai, however, the web sites were once again blocked.[66]

While access to sites of groups such as Human Rights Watch/Asia and Human Rights in China are routinely blocked, Beijing has also shown a willingness to block prominent nondissident sites for hosting "anti-China" material. After *Tiananmen,* a magazine published by exiled Tiananmen student leader Wang Dan, was posted on a Stanford University server, Beijing responded by blocking all Stanford-hosted sites.[67] The California Institute of Technology was allegedly pressured by Beijing to remove a Falungong site from its servers. When the request was rejected, access within China to the university's entire web site was blocked for several months.[68]

One site that has been repeatedly blocked by the Beijing authorities is that of *China News Digest* (CND), a volunteer group of overseas Chinese clearly unsympathetic to the regime. An account of how the CND site was blocked is offered by its founder, Wei Lin:

> Our site was among the first batch of 100 or so web sites being blocked. Some time after that, there were reports about some sites actually being removed from the Chinese government's blacklist. Earlier this year [1998], readers in China were able to access our site because we changed our ISP and consequently the IP address for our web server changed. This lasted until 5 June 1998, then the China traffic suddenly dropped. We had linked to the June 4th Beijing Massacre photo archive from our top page, which had increased our hits from 2000/day to 4000/day. Our ISP was nice enough to let us change the web server IP once again. But our server log showed that every time we changed the IP address, the traffic from China would last for a day or two during the month of June before the new IP is blocked again. We can pretty much assume that they are using the latest software on the routers when available and applicable. They do not have any technology lag other than possibly the export restriction from the U.S. government regarding encryp-

---

[66]This is done even for some relatively routine visits by less-senior foreign officials, reporters, and other guests.

[67]Scott Savitt, "China's Internet Revolution," *Asian Wall Street Journal,* December 21, 1999, p. 10.

[68]"Top U.S. Institute Won't Bow to Dictatorship," Central News Agency, February 20, 2000.

tion. They have implemented classless interdomain routing (CIDR). Commercial ISPs connect to the state-owned backbones, ChinaNet and ChinaGBN, [and] many of them implement private networks and users access the Internet via proxies. The method used to block sites outside of China is, because the government allows only a few backbone networks to exist and to have their own international links, to put the access-control list on all border routers (which appear to be CISCO routers). This is very efficient because it blocks based on [the] packet source's IPs and normally those sites cannot change their IPs easily. Based on our experience in early June 1998, CERNET and CASNET blocked us first, ChinaNet a day later, GBN some time later.

Wei closed by noting that since ChinaNet and ChinaGBN are both part of the merged information superministry, the sequenced delays in adding the blocking would suggest that, administratively, their networks are not fully merged.

Despite these types of efforts to block sensitive sites, however, CND is at the forefront of the promotion of one of the most potent weapons used in fighting web-site blocking: proxy servers.[69] The CND web site contains a guide to using proxy servers to circumvent firewalls or Internet censorship, as well as a call for volunteers to provide CND with additional proxy services.[70] A CND web page (http://proxy.cnd.org/) contains detailed instruction on the use of proxies, including various techniques for configuring web browsers. For its users, CND currently provides four official proxies, each of which uses a dynamic IP address:

1. http://cnd-d.cnd.org:8000/http://www.cnd.org

2. http://proxy2.cnd.org:8000

3. http://proxy3.cnd.org:8000

4. http://anon.free.anonymizer.com/http://www.cnd.org

---

[69]A proxy server sits between a client application, such as a web browser, and a real server. Proxy servers are also used to improve performance or filter requests. For more information, see www.webopedia.com.

[70]See www.cnd.org. CND seeks volunteers who have system privileges on UNIX servers or on PCs running Linux, and it notes that individuals who work at universities and have cable modems or DSL service are particularly desirable volunteers.

The fourth proxy exploits one of the most popular "anonymizer" web services, allowing users to surf the web in relative privacy. The CND site also points users to other proxy lists, available in abundance by searching portals such as Yahoo! or Webcrawler for the term *proxies.*

New York–based Human Rights in China (HRIC) has also published an article describing the use of proxy servers to access blocked web sites in its quarterly *China Rights Forum.*[71] Available evidence is fragmentary but suggests that many mainland Internet users are capable of accessing a variety of blocked sites. Xiao Qiang, executive director of HRIC, says that many sites that are blocked by PRC authorities, including those of HRIC and CND, still receive numerous visits and e-mail messages from web surfers on the mainland.[72] Statistics provided by CND on its web site confirm that some mainland users can access the site even though it is blocked.[73] CND also advises readers who have difficulty accessing their site from within China that they can receive CND publications via e-mail, using accounts at www.usa.net, www.hotmail.com, www.yahoo.com, and other free web-based e-mail services.[74] In addition, Zhang Weiguo recommends that PRC readers use proxy servers to access his *New Century Net* site, which the Chinese authorities began blocking about a year after its establishment. According to Zhang, as of 2000, new proxy addresses could generally be used for about two months before they were discovered and blocked, but this is reportedly no longer the case, as the Chinese authorities have redoubled their efforts to block access to popular proxy servers (this is discussed in further detail below).

As a result of these efforts and the gaps in the implementation of blocking on the Chinese end of the system, most sensitive sites are

---

[71]"Proxy Servers," *China Rights Forum,* Human Rights in China, fall 1998.

[72]Erik Eckholm, "China Cracks Down on Dissent in Cyberspace," *New York Times,* December 31, 1997.

[73]According to CND's access statistics for Friday, June 9, 2000, for example, at least six mainland Chinese users (defined as visitors with .cn domain names) accessed the main CND page, and at least 16 reached one of CND's two mirror sites, presumably by using proxy servers. A mirror site is a replica of an already existing site, used to reduce network traffic (hits on a server) or improve the availability of the original site. For more information on mirror sites, see www.webopedia.com.

[74]"Some Special Notes for Network Users in CN Domain (Mainland China) or Accessing to the Internet from Behind Firewall (Need a Proxy?)," CND web site.

available to the Chinese population, or at least to those Internet users who are willing to devote some time and effort to accessing them. Among these, a number of avowedly pro-democracy web sites, human-rights web sites, and Tibet-related web sites, including that of the Tibet government-in-exile, continue to be accessible even without the use of proxy servers.[75] The authors' own limited empirical studies in Internet cafes in major Chinese cities, usually involving 100 to 200 political and media sites, reveal little consistent blocking of even the most sensitive sites. More surprising, the vast majority of terminals in these cafes, including those populated exclusively by locals, were preconfigured with the necessary proxy servers in Australia or elsewhere to circumvent the "Great Firewall."

Beginning in early 2001, however, the Chinese authorities stepped up their efforts to block access to well-known proxy servers and privacy protection sites. Companies offering privacy protection services (e.g., Safeweb, SilentSurf, and Anonymizer) are now engaged in a daily game of measure and countermeasure with the authorities in China and other countries that try to restrict access to particular web sites.[76] For example, the Safeweb site was blocked by Chinese authorities beginning in late February 2001.[77] At around the same time, SafeWeb received funding from the International Broadcasting Bureau, which oversees the Voice of America, to enhance the services it offers to Chinese users who want to access blocked web sites.[78] The result was a peer-to-peer application, Triangle Boy, released in April 2001, aimed at enabling Chinese Internet users to access all websites blocked by the authorities.

---

[75]U.S. Department of State, *Human Rights Report for 1999.*

[76]See for example, Jennifer 8. Lee, "Punching Holes in Internet Walls," *New York Times*, April 26, 2001.

[77]Recent research suggests that this version of Safeweb, which was taken out of service in November 2001, may not have provided the degree of anonymity it promised its users. Researchers from Boston University and the Privacy Foundation assert that Safeweb's security vulnerabilities would "allow adversaries to turn Safeweb into a weapon against its users, inflicting more damage on them than would have been possible if they had never relied on Safeweb technology." See David Martin and Andrew Schulman, "Deanonymizing Users of the Safeweb Anonymizing Service," February 11, 2002, p. 1. The full report is available online at http://www.cs.bu.edu/techreports/pdf/2002-003- deanonymizing-safeweb.pdf.

[78]Jennifer 8. Lee, "U.S. May Help Chinese Evade Net Censorship," *New York Times*, August 30, 2001.

Despite claims that Safeweb's Triangle Boy system would allow Chinese Internet users unfettered access to blocked sites,[79] however, the Chinese authorities reportedly have had a relatively easy time blocking the system. Computer-savvy Internet users have reported that they are unable to access Safeweb's Triangle Boy servers, and the authors were also unable to use the system on two recent trips to the PRC. One major problem with the current system is that of communicating the IP addresses of Triangle Boy servers to people in China without the authorities also finding out and quickly blocking access to the sites. Currently, Safeweb sends out an e-mail containing the information in response to user requests. Personnel from the Ministry of Public Security's Computer Monitoring and Supervision Bureau simply request the information through e-mail as often as necessary and block the IP address on the routing table.[80] After only a few months, Beijing's blocking of the addresses caused an 80 percent decline in the use of Triangle Boy by Chinese web surfers, according to a Western media report.[81] "Now it is basically impossible to use it," laments an article posted on a Chinese-language web site dedicated to promoting freedom of speech on the Chinese Internet.[82] This article and other postings on Chinese-language web sites indicate that since the Chinese authorities intensified their efforts to block access to Safeweb's Triangle Boy network, some mainland Internet users have turned to another peer-to-peer application, the Chinese version of Freenet, which can be used not only to search for and exchange documents, but also to access blocked web sites. Nevertheless, it appears that, for the moment at least, Beijing has the up-

---

[79]See Safeweb's "White Paper: Triangle Boy Network," available on the Safeweb web site, www.safeweb.com. The paper asserts: "Triangle Boy is our answer to Internet censorship. Triangle Boy defeats all attempts to prevent users from accessing sites on the Internet." Safeweb expects that it will become more and more difficult for the Chinese security services to block access to its network over time as the number of Triangle Boy machines increases. The use of dynamic IP addresses by some Triangle Boy machines, according to Safeweb, "makes the censors' task even more daunting and the likelihood of success even slimmer." For now, it appears that Beijing has the upper hand in its battle with Safeweb, but as these points illustrate, this could change in time.

[80]Interviews, Western computer technicians, January 2002.

[81]Pamela Yatsko, "China's Web Censors Win One—For Now," *Fortune*, December 24, 2001.

[82]*"Guanyu Ziyouwang"* [About Freenet], December 11, 2001, available at www.internetfreedom.org/gb/articles/1042.html.

per hand in the web-site blocking battle. In the words of one activist who recently visited the mainland, "The authorities have become much better at finding and blocking proxies in China; I was there for eight days and experienced eight days of Internet blackout."[83]

In sharp contrast to web sites, e-mail and e-mail publications are difficult to block, although the PRC government attempts to do so at times by blocking all e-mail from overseas ISPs used by dissident groups. The Voice of America Chinese-language e-mail news server was blocked beginning in April 1999, except for a brief period in July of that year. Beijing has responded to the spamming of *Xiao Cankao* by blocking all e-mails from the ISP from which the messages originate.[84] As noted earlier, at the height of the summer 1999 crackdown on Falungong, the government ordered an ISP suspected of being a conduit for the group's international coordination activities to suspend e-mail service for two days. In the run-up to June 4, 2000, the Ministry of Public Security reportedly sought to slow down or shut down China's free and anonymous e-mail sites, 163.net and 263.net. As noted earlier, the owners of 163.net, Liang Liwei and Leng Wanbao, said that they could neither send nor receive e-mail beginning on June 1.

Despite these government efforts, opponents of the regime have enjoyed significant success in overcoming e-mail blocking. Indeed, as argued earlier in this report, the spamming campaigns of *Xiao Cankao* and *Da Cankao,* originating from a different e-mail address every time, are among the most successful dissident Internet strategies. Even the CND listserv e-mails, which originate from the same site, have enjoyed relatively easy access, due in part to the advice offered by CND on how to overcome e-mail blocking. In particular, its staff recommends that mainland users obtain free Web-based e-mail accounts, such as www.usa.net, www.hotmail.com, www.mailcity. com, www.yahoo.com, or www.rocketmail.com. After sending mail/ subscribe messages to CND from these sites, users are free to check

---

[83]Interview, Chinese activist, March 2002. As peer-to-peer applications become more widely used in China, however, the advantage now enjoyed by the authorities may shift back to those Internet users who are interested in viewing banned web sites.

[84]Frank Langfitt, "Taking Dissent Online in China; E-mail: In the Age of the Internet, Chinese Leaders Are Finding It Harder to Contain Free Speech," *The Baltimore Sun,* May 11, 1999, p. 2A.

their e-mail on the web, with no fear of receiving e-mails that must traverse Chinese government-controlled routers.

Hacking. There is some evidence to suggest that the Chinese government or elements within it have engaged in hacking of dissident and antiregime computer systems outside of China. Given the inherently indeterminate nature of the source of most computer network intrusions, it is often difficult if not impossible to establish official culpability for hacking attacks without additional evidence. Governments, usually by design, can therefore claim a reasonable measure of plausible deniability in these cases. The Chinese-origin hacking attacks that occurred against Taiwan in August 1999 and against Japan in February 2000 are examples of incidents in which government culpability, either limited or complete, is difficult to determine solely on the basis of the intrusion data.

Stronger evidence exists to support the conclusion that the Chinese government or elements within it were responsible for one or more of the China-origin network attacks against computer systems maintained by practitioners of Falungong in the United States, Australia, Canada, and the United Kingdom. After the exposure of the role of certain Chinese security agencies in the attacks, the later, more sophisticated intrusions were believed to have been carried out by cut-outs, making it more difficult to ascertain the extent of government involvement. This was especially true of the attacks that occurred in winter and spring 2000.

Summer 1999. In mid-July 1999, the Chinese government authorities began a nationwide crackdown on the Falungong organization, claiming that it was a "dangerous cult." News of the crackdown spread quickly, due in large measure to the organization's extensive use of advanced information technologies and its network of Internet sites around the globe. These sites provided real-time accounts of crackdowns in some Chinese cities, based on e-mails and other communications from Falungong members. As the story was gradually picked up by the global media, these sites, many of which were shoestring operations run by group members, understandably began to strain under the increased hits they received. While this slowdown in service was an expected consequence of worldwide attention, some of the sites began to suffer from anomalous crashes. When the system administrators of these servers examined the situation in detail, some realized that their networks were suffering from a sophisticated series of computer network attacks. The July 1999 attacks against Falungong sites in four countries (one in Britain, two in Canada, one in Australia, and two in the United States) bear greater scrutiny.

The evidence of a Chinese government-directed information operation against Falungong is strongest in the U.S. case. On July 14, 1999, Falungong practitioner Bob McWee of Middletown, MD, established www.falunusa.net, with the express purpose of mirroring the files of existing Falungong sites in Canada (www.falundafa.ca and www. minghui.ca) and the United States (www.falundada.org).[85] On July 20, 1999, the two Canadian sites began to suffer a degradation of network performance, because of Chinese-origin hacking attacks. As a result, they began re-routing connection requests to their mirror site, FalunUSA. Between July 21 and 23, the U.S. site began to have similar difficulties. Specifically, it was suffering from a type of attack known generally as a denial-of-service attack, in which the target machine is flooded with incomplete requests for data and eventually succumbs to the attack by crashing. Backtracking a similar attack on July 27, 1999, revealed the source IP address of the attack to be 202.106.133.101, an Internet address in China. Examination of the Asia-Pacific Network Information Center (APNIC)[86] database entry for this address revealed the ownership information shown in Figure 1.

The name of the organization, "Information Service Center of XinAn Beijing," sounded innocuous enough, but the street address told a very different story. The address, #14 East Chang'an Street (listed in Figure 1 in transliteration as "Dong Chang An Jie 14") in Beijing, is that of the Ministry of Public Security, China's internal security service—the organization most embarrassed by the unexpected appearance of thousands of Falungong practitioners outside the

---

[85]Svensson, "China Sect."

[86]APNIC is the Internet registry organization for the Asia-Pacific region. For more information on APNIC, see http://www.apnic.org.

Inetnum:	202.106.133.0 - 202.106.133.255
Netname:	ISCXA
Descr:	Information Service Center of XinAn Beijing
Country:	CN
Admin-c:	WH42-AP
Tech-c:	HJ36-AP
Changed:	suny@publicf.nta.net.cn 19990716
Source:	APNIC
Person:	Wang Huilin
Address:	Dong Chang An Jie 14 Beijing 100741
Phone:	+86-10-65203827
Fax-no:	+86-10-65203582
Nic-hdl:	WH42-AP
Changed:	suny@publicf.bta.net.cn 19990716
Source:	APNIC
Person:	He Jian
Address:	Dong Chang An Jie 14 Beijing 100741
Phone:	+86-10-65203789
Fax-no:	+86-10-65203582
Nic-hdl:	HJ36-AP
Changed:	suny@publicf.bta.net.cn 19990716
Source:	APNIC

Figure 1—Original APNIC Database Entry

usually not sufficient to identify whether the true source of an attack is the organization in question or simply a third party that has hacked into the MPS network and used it as a base to launch attacks. Four crucial pieces of evidence, however, strongly suggest that the MPS was the real culprit in the attacks against Falungong sites. First, the network had been established shortly before the information operations began and was divorced from other explicitly identified MPS networks in other parts of Chinese cyberspace, such as the network belonging to the Public Security Bureau of Jilin Province (202.98.12.64 – 202.98.12.71) or the Shaanxi Public Security Bureau

Computer Monitoring Unit (202.100.14.128 – 202.100.14.143). Second, the name of the organization in the database—Information Service Center—suggests an intent to deceive outsiders about its actual affiliation. Third, at least one Western media source claimed to have called the telephone numbers listed in Figure 1 and was told by the person answering the phone that the numbers belonged to the Ministry of Public Security.[87] A later call by the same news organization to the telephone operator at the ministry confirmed that the numbers belonged to the MPS Internet Monitoring Bureau.[88] The fourth and most telling piece of evidence resulted directly from the impending exposure in the Western media of the network's governmental affiliation. As the reporters were filing the stories with their editors in the United States, the information in the APNIC database was being altered—in particular, the damning street address of the owner of the network was being deleted and replaced (see Figure 2).

If the ministry's network had itself been the victim of an attack and was thus wrongly accused as the perpetrator of the attacks on the Falungong site in the United States, why go to the trouble of changing the database information to an address other than MPS headquarters? And was it a coincidence that the network information was changed on the eve of an exposé in a major Western newspaper of the MPS's alleged role in the attack? The evidence cited earlier, along with this attempt to disguise the true owner of the network, strongly suggests that the perpetrator was caught with its "hand in the cookie jar."

Of course, the fact that the attacks might have originated from an MPS network does not automatically imply that they were sanctioned by the ministry leadership or their superiors in the senior party leadership. One possibility that must be considered is that the attack was carried out by a "rogue element" within the MPS, without approval from anyone. After the exposure of a rogue's efforts, a natural reaction would be to cover up the network's ministry affiliation by changing the APNIC data. One might question whether the ministry would be able to find the perpetrator, conduct an investigation of his

---

[87]Svensson, "China Sect."
[88]Ibid.

Inetnum:	202.106.133.0 - 202.106.133.255
Netname:	ISCXA
Descr:	Information Service Center of XinAn Beijing
Country:	CN
Admin-c:	HJ36-AP
Tech-c:	HJ36-AP
Changed:	suny@publicf.nta.net.cn 19990716
Changed:	suny@publicf.nta.net.cn 19990729
Source:	APNIC
Person:	He Jian
Address:	Zheng Yi Lu 6 Dong Cheng District Beijing 100741
Phone:	+86-10-68765432
Fax-no:	+86-10-68765432
Nic-hdl:	HJ36-AP
Changed:	suny@publicf.bta.net.cn 19990716
Changed:	suny@publicf.nta.net.cn 19990729
Source:	APNIC

Figure 2—Altered APNIC Database Entry (July 29, 1999)

actions, and implement a technical fix so quickly, but as improbable as that seems, it is not impossible.

One final footnote to the July 27, 1999, attack against FalunUSA.net: The manner in which the MPS allegedly brought down the site contains a fascinating twist. The denial-of-service attack was a classic "SYN flood" attack and appears to have been designed to make it appear as if Falungong was conducting information operations against the U.S. Department of Transportation (DOT).[89] In the July attack,

---

[89]Any successful connection between two servers on the Internet requires a three-way "handshake" before information can be exchanged. First, Machine A sends a SYN to Machine B, which responds to Machine A with a SYN-ACK. Machine A then closes the loop by sending Machine B an ACK. The success of this exchange requires that all of the packets contain correct address information; otherwise, they will go to the wrong places. A SYN flood exploits this dynamic. In such an attack, Machine A sends a SYN with an incorrect return address to Machine B, which logically responds by sending its SYN-ACK not to Machine A but to Machine C. Since both Machine B and Machine C

the MPS network sent a SYN to the FalunUSA site with an incorrect return address, namely, a server controlled by DOT. A network engineer at DOT contacted Bob McWee and the operators of the other Falungong sites to find out why www.falundafa.org, www.falunUSA.net, and www.falundafa.ca were sending unauthorized packets to a DOT server, according to Everett Dowd, deputy director of telecommunications in the DOT Information Technology Operations office.[90]

Why, out of the millions of possible IP addresses, did the MPS choose an address belonging to DOT? One plausible hypothesis is that the perpetrator wanted a "two-fer": crash the Falungong site, but also make it look as if the Falungong site was engaged in information operations against a U.S. government site. At the time of the attack, the entire Chinese governmental propaganda apparatus was in high gear, branding Falungong a "dangerous cult" and a "terrorist organization." What better way to demonize Falungong than to make it appear that the organization was hacking sites run by the U.S. government? Indeed, system administrators at DOT initially thought they were under a different type of denial-of-service attack (a SYN-ACK flood) from the Falungong site, since all they could see on their end was a series of SYN-ACK requests entering their system from FalunUSA.net for no apparent reason. Only later did the DOT personnel realize that the Falungong site had simply been the unwitting accomplice of a third party.

Attacks on Falungong sites in England and Australia during late summer 1999 bear some interesting similarities to the intrusions in the United States, particularly with regard to the source IP addresses of the perpetrators. The U.K. Falungong web site (http://www.yuanming.org.uk) was set up on July 20, 1999, by Zhu Bao, a Falungong practitioner living in Dublin, Ireland.[91] By July 23–24, 1999, the site had come under continuous attack from China-origin IP addresses. At the beginning of the attacks, the intruders disabled the

---

have a limited number of slots in their buffers for these sorts of unanswered queries, they both eventually suffer from buffer overflow and crash.

[90]Associated Press, August 6, 1999.

[91]Jonathan Dube, "China Ate My Web Site," ABCNEWS.com, August 6, 1999.

server.[92] Later, they deleted all the original files and replaced them with the text of an article from the *Xinhua* News Agency entitled "The Person and Affairs of Li Hongzhi," falsely listing the author of the article as a member of the "Falungong Research Society." The article says that Li

> is not the "highest Buddha" who brings salvation to suffering people, but an evil person who has had an extremely disastrous effect upon society. Li is not bringing salvation to practitioners, but is in fact leading them to a disastrous and miserable end, and Falungong is doing enormous harm to both the mental and physical health of people.

Falungong's U.K.-based service provider (NetScan, www.netscan.co.uk) confirmed that the intruders had obtained their root password.

In a separate attack, Li Shao of Nottingham publicly reported on July 26, 1999, that his Falungong site was attacked by hackers operating from a Chinese IP address.[93] Falungong sources claim that the British police linked the address to the Information Service Center of XinAn in Beijing, discussed above, but no independent confirmation was possible.[94]

In Canada, two Falungong sites (www.minghui.ca and www.falundafa.ca) were attacked by hackers, and both eventually succumbed. The ISPs for these sites, Bestnet Internet of Hamilton, Ontario, and Nebula Internet Services of Burlington, Ontario, reported that their networks were attacked on July 30, 1999, by Chinese government servers because they hosted sites run by Canadian followers of Falungong, including Jason Xiao, the system administrator of www.falundafa.ca.[95] According to the director of Bestnet Internet, Eric Weigel, the hack attempts originated with "Chinese government offices in Beijing." Weigel stated that the specific originating addresses belonged to the Beijing Application Institute for Information

---

[92]The details of this attack are derived from Falungong, "Report," p. 23.

[93]Svensson, "China Sect."

[94]Falungong, "Report," p. 87.

[95]Peter Goodspeed, "Falung Gong, Beijing Wage War over Internet," *National Post,* November 2, 1999.

Technology (BAIIT) and the Information Center of XinAn Beijing.[96] No IP addresses were furnished by the newspaper accounts, but BAIIT's networks can be found between 203.93.160.0 and 203.93.160.255. Possible government connections are suggested by the P.O. box mailing address provided by BAIIT in the APNIC database, as P.O. boxes are often used in lieu of street addresses by Chinese government and military hosts. By contrast, the government affiliations of the Information Center of XinAn Beijing are much clearer, as discussed in greater detail earlier in this chapter.

Nebula Internet Services reported that the same sites had attempted to crash its servers, using similar types of attacks. According to Nebula representatives, the assault went on for more than a month, coinciding with the timetable of the government crackdown on the sect. Unlike Bestnet, which had more-advanced equipment and was able to withstand the attacks with little loss of service, Nebula's systems were crippled by the hackers, and the company was forced to shut off its service. The owner of two Canadian Falungong sites (perhaps the same sites discussed above), Jillian Ye of Toronto, claimed that her sites had been under attack every day for several months and that the problems had gotten progressively worse until she finally moved the sites to a more secure server.[97]

Fewer similarities exist between the attacks described above and those against Falungong servers based in Australia, but the timing of the Australian attacks (in late summer 1999 and mid-spring 2000) coincides to a significant degree with attacks in other countries. An Australian practitioner of Falungong established a Falungong mirror site (http://falundafa. au.cd) in March 1997 on a Windows NT server.[98] On September 6, 1999, computer attacks originating from a Chinese IP address forced this site to shut down.[99] The victims reported to the police that the intruders tampered with their e-mail system. The system administrator of the site noticed that the infiltrators were able to manipulate the cursor on their screen, which suggests that the attackers were using a hacker tool known as Back

---

[96]Oscar Cisneros, "ISPs Accuse China of Infowar," *Wired News*, July 30, 1999.

[97]Svensson, "China Sect."

[98]Interview, Falungong practitioner, June 2000.

[99]"Falungong Hot on Jiang's Trail," *Agence France Presse*, September 7, 1999.

Orifice to penetrate the site. Beginning in September 1999, Australian police undertook constant monitoring of the site.

Spring 2000. The first of the renewed attacks against Falungong servers occurred on March 11, 2000, coinciding with the meetings of the National People's Congress in Beijing. The hack, which used a denial-of-service technique known as a "smurf" attack, brought down the main server in Canada (www.minghui.ca), as well as three mirror sites (www.falundafa.ca, www.falundafa.org, and www. minghui.org).[100] Since smurf attacks are quite effective in masking the identity of the attacker, no useful source information could be gained from the logs of the intrusions.

Attacks on Falungong servers reached a crescendo in mid-April 2000, when five sites—three in the United States (www.falunUSA.net, www.falundafa.org, www.truewisdom.net) and two in Canada (www. minghui.ca and www.falundafa.ca)—were smurf-attacked simultaneously.[101] The timing of the attacks coincided with two sensitive political events: (1) the impending vote in the United Nations Human Rights Commission on a UN resolution condemning Chinese human-rights abuses, including persecution of Falungong; and (2) the one-year anniversary of the April 25, 1999, gathering of Falungong practitioners outside the central leadership compound in Beijing.

---

[100]Smurf attacks employ a two-step procedure. First, hackers scan the Internet for vulnerable servers or host computers. Ideal target systems have relatively wide bandwidth and few IP addresses, characteristics found in servers operated by universities (.edu) and nonprofit organizations (.org). The networks of these servers are often composed of subnetworks. Usually, a request sent to the main IP address is answered by every computer on the local network. In other words, if the local network has 40 subnetted computers, one request will result in 40 replies. These types of servers can be used as "Internet request amplifiers" or "slaves" for a smurf attack. Hackers will assemble large numbers of these slaves for an impending attack, hoping to direct all of their bandwidth toward a single target server.

In the second step, hackers issue the signal to the slaves. Attackers forge a ping command that appears to be coming from the target computer. For every fake ("spoofed") ping they send, the victim is flooded with many (40, in our example) replies. A dial-up user with 28.8 kbps of bandwidth exploiting this technique on our illustrative network could generate (28.8×40) or 1152.0 kbps of traffic, about 2/3 of a T1 link. The smurf attacks that brought down eBay and Yahoo! used much larger sets of networks.

[101]"Web Sites of Falungong Hit," *Agence Prance Presse*, April 14, 2000.

Falungong system administrators received a variety of warnings about the impending attack. Around April 6, Falungong received an e-mail warning that the Public Security Bureau had paid two network security companies to hack the group's sites abroad. After the first wave of attacks, Falungong system administrator Li Yuan received an anonymous tip on April 12 confirming the situation. "We received an anonymous e-mail from a Chinese computer expert on April 12 warning us that the police computer security bureau had offered to pay a computer company money to hack into our sites," said Yuan.

According to the Maryland-based system administrator for FalunUSA, the attacks themselves began around April 9 or 10. The intruders attacked the IP addresses of the sites, not the domain names, and likely got into the system using security holes in the ftp command. Once inside, the attackers replaced most of the original network command files (e.g., *ls*, *df*, and *find*) with versions of these files that contained "trojan horses" for later penetration. The system administrator reports that after he discovered and dismantled the hackers' efforts, intruders attempted to log on to his server, using ftp and SSH commands, but these probes were rebuffed.

In Australia, the attacks started again between March and May 2000, with the most serious attack coming on May 22. The Australian server was crashed by hackers around 3 a.m. on May 22, rebooted the next morning, and hacked again one hour later. It was not rebooted a second time until 7 or 8 p.m. Logs of these attacks and the addresses of the attacking sites were unavailable for analysis, but the Australian system administrator said that the intruders used an exploit known as IISATTACK, and their IP addresses could be traced to Hong Kong, England, and the United States. The system administrator asserted that the attacks in 2000 were far more sophisticated than those in 1999, and the attackers were able to easily exploit the server's remote logins, which were later disabled by its owners.

Monitoring and Filtering. Foreign visitors to China and domestic dissidents have long been aware that the Chinese government is engaged in widespread monitoring of communications. According to the 2000 State Department China human-rights report:

> [The Chinese] authorities often monitor telephone conversations, fax transmissions, e-mail, and Internet communications of citizens,

foreign visitors, businessmen, diplomats, and journalists, as well as dissidents, activists, and others.[102]

The extent of this monitoring, however, is frequently overstated, as the sheer scale of the necessary effort is beyond the resources of the security apparatus. This is especially true of electronic communication.[103] Members of the security apparatus suggested in interviews that they recognize the technical difficulties—or, rather, the impossibility—of wide-scale e-mail monitoring, regardless of encryption. While research in keyword searching applications continues, even its advocates realize that a network system would grind to a halt if keyword searches were attempted on a nationwide or even a regional basis, given the enormous volume of electronic communication.[104] Public security sources confirm that selective, often *post hoc*, monitoring, combined with traditional surveillance methods, is a preferable and far more effective strategy.[105]

Fragmentary evidence exists to support the notion that the security services possess and are actively developing limited monitoring and filtering capabilities. The anonymous author of an article entitled "China's Main Methods of Supervising and Controlling the Net and Countermeasures," which was recently published on a Chinese-language dissident web site, asserts that there are two types of methods used by the Chinese authorities to monitor and control e-mail communications: filtering software (*guolu ruanjian*) and selective examination (*choucha*) of users' electronic mailboxes.[106] This is confirmed by a number of other sources. In a published interview, a computer engineer claiming to be employed by the Public Security

---

[102]U.S. Department of State, *China Country Report on Human Rights Practices, 2000*, February 23, 2001, p. 15.

[103]Recently, the apparently modest results of U.S. efforts to track terrorists through the Internet have illustrated the difficulties of conducting online surveillance against users who seek to evade detection by communicating with each other via anonymous e-mail accounts accessed at Internet cafes, sometimes using strong encryption. See, for example, Susan Stellin, "Terror's Confounding Online Trail," *New York Times*, March 28, 2002.

[104]Interviews, Western executives, January 2000.

[105]Interviews, PRC officials, January 2000.

[106]*Zhongguo dui wangluo de zhuyao jiankong fangfa he duice* [China's Main Methods of Supervising and Controlling the Net and Countermeasures], www. internetfreedom.org/gb/articles/1012.html.

Bureau asserted that his organization monitors information aimed at "undermining the unity and sovereignty of China" (i.e., references to Tibetan independence or the Taiwan question), communications that attempt to propagate new religions, and dissident publications by filtering for selected keywords.[107] Human Rights Watch reported that in May 1998, the Ministry of Labor and Social Security installed monitoring devices at the facilities of ISPs that can track individual e-mail accounts.[108] In January 2000, Liu Ming, the younger sister of student leader Liu Gang, wrote an indictment against the Changchun City Public Security Bureau, which was then disseminated abroad via the Internet by Leng Wanbao, a noted dissident in Changchun City, Jilin Province. Leng was picked up and interrogated for three hours by the police, who knew about the activity immediately.[109] A Heilongjiang Harbin University student, Zhang Ji, was arrested for disseminating "reactionary information via the Internet." He was alleged to have been sending e-mail messages to Falungong web sites in the United States and Canada, as well as downloading information from those web sites and relaying it to fellow Falungong practitioners. Falungong sources in the United States believe that police copied the e-mail addresses of Falungong net users, and their e-mail passwords were obtained from mainland ISPs. As a result, the U.S. Falungong sources believe that all e-mail from Falungong practitioners that passes through 163.net and 263.net is now monitored by the Chinese government. Falungong practitioners in China also claim that the Public Security Bureau has intensified its efforts to identify Internet users who try to access the group's overseas web sites.[110] Finally, the Committee to Protect Journalists asserted in its 1999 report that the Ministry of State Security has an entire department devoted to tracking dissidents and their writings on the Internet.[111]

---

[107] Geremie Barme and Sang Ye, "The Great Firewall of China," *Wired 5.06*, June 1997.

[108] U.S. Department of State, *China Country Report on Human Rights Practices, 1999*.

[109] "Report on PRC Controlling Dissidents' E-mail," Hong Kong Information Centre for Human Rights and Democratic Movement in China, January 19, 2000, in *FBIS*, January 19, 2000.

[110] Craig S. Smith, "Sect Clings to the Web in the Face of Beijing's Ban," *New York Times*, July 5, 2001.

[111] "State Tracks Dissidents Online," Associated Press, March 24, 2000.

Chinese Internet users have responded to these efforts with a variety of protective countermeasures.[112] One of the simplest, but most effective, consists of logging onto the Internet anonymously at one of the Internet cafes that are by now ubiquitous in many Chinese cities. Chinese-language Internet postings advise mainland users on appropriate security measures for covering their tracks at Internet cafes.[113] In some cases, it is also possible to exploit gaps in the monitoring technology. For instance, some net companies use software to identify occurrences of a leader's name in online postings so they can remove any unfavorable comments, but chat room visitors reportedly dodge that restriction by putting a space between characters or using nicknames.[114] Other users protect themselves by discretion. Before June 4, 2000, the discussion of Tiananmen issues quieted down significantly. Frequent visitors to popular chat rooms reportedly posted warnings asking fellow chat room visitors to avoid leaving sensitive messages about June 4 on the Internet. On June 1, some messages in chat rooms claimed that Internet portals had received warnings from police to delete messages about the 1989 movement, as well as messages about the recent murder of Beijing University student Qiu Qingfeng.[115]

Finally, users have sought safety in numbers. After Chen Shui-bian's inauguration speech in May 2000, Chinese BBS were overflowing with critiques and commentaries, many of which disagreed with the PRC government's policy and conduct. The BBS on the *People's Daily* web site became a particularly intense hotbed of comment, overwhelming the ability of the censors to deal with it, though they

---

[112]See, for example, Shi Lei, "*Xinxi bailinqiang: tupo zhonggong wangluo dianzi youjian fengsuo (zhiyi),*" [The Information Berlin Wall: Breaking the Chinese Communist Party's Net and E-mail Blockade (Part One)], November 2, 2001; and Shi Lei, "*Zhonggong ruhe guolu he jiecha wangluo dianzi youjian: tupo zhonggong wangluo dianzi youjian fengsuo (zhi er)*" [How the Chinese Communist Party Filters and Monitors the Net and E-mail (Part Two)], November 2, 2001, available on the web site of the Home for Global Internet Freedom at http://www.internetfreedom. org/gb/articles/997.html.

[113]"*Wangba shangwang de yixie anquan wenti*" [Some Security Problems of Going On-Line at Internet Cafes], internetfreedom.org/gb/articles/979.html.

[114]Julie Schmit and Paul Wiseman, "Surfing the Dragon: Web Surfers Find Cracks in Wall of Official China," *USA Today*, March 15, 2000, p. 01B.

[115]Josephine Ma, "Cyber-Crackdown Fails to Silence Protesters," *South China Morning Post*, June 2, 2000.

were able to prevent the posting of the text of Chen's speech.[116] This was an interesting choice of priorities, highlighting the extent to which the government seeks to deny the population access to primary sources of information that would allow them to form opinions other than those generated by the propaganda apparatus.

Propaganda, Denial, Deception, and Disinformation. The deceptive practices of the Chinese government on the Internet take a number of forms. The first is *denial,* in the strict sense of the word. In October 1998, the PRC government-controlled Chinese Society for Human Rights, which represents China in its expanding human-rights dialogue with other countries, launched a web site (www.humanrights-china.org) to promote Beijing's official line on the subject and deny the accounts of Chinese human-rights abuses alleged by nongovernmental organizations. The site contains government documents in Chinese and English, including articles from the state-run media, legislation, and lectures from a symposium on human rights that Beijing hosted.[117] (U.S.-based "hacktivists" successfully penetrated this site in late October 1998, replacing its front page with an impassioned attack on the Chinese government's repressive policies and a trenchant criticism of the site's poor network security.[118])

The second deceptive tactic is *passive disinformation about third parties.* For example, as part of the campaign against Falungong, the Chinese Academy of Social Sciences (CASS) has established a web page to combat cults and expose their evil intentions and activities. The web site is divided into six topics, including theories of Marxism, atheism, and materialism, as well as analysis and exposure of the Falungong cult and a general survey of cults in other countries. Its content is similar to the overall anti-Falungong propaganda campaign that has been waged relentlessly by Beijing since August 1999.

---

[116]Michael Dorgan, "Chinese Censors Losing Online Race," *San Jose Mercury News,* May 22, 2000.

[117]"China: We're Only Human," Reuters, October 26, 1998.

[118]For an account of the hack, see "China Cyber-Cops Partially Block Hacked Web Site," Reuters, October 29, 1998. For the Chinese response to the attack, see "Hacker Attacks Society for Human Rights Studies Web Site," *Xinhua,* October 29, 1998, in FBIS, October 29, 1998.

CASS claims that as of May 10, 2000, the site had received over 500,000 hits.[119]

The third deceptive tactic is *active disinformation,* including harassment and character assassination. One victim of such harassment, Frank Siqing Lu, claimed that after the suppression of the CDP in late November 1998, public security offices used the names of local dissidents to page him and leave return numbers that were either nonexistent or numbers for hotels, karaoke bars, and hospitals. Public security officials also allegedly bombarded his fax machine with large numbers of blank pages.[120]

Similar tactics are reported by Falungong members both in China and abroad. There is some evidence to suggest, for instance, that individuals have been redirecting e-mails from China to send fake articles by Li Hongzhi to Internet users in China, using a flaw in Internet mail protocols that is easily unmasked. On February 28, 2002, a Falungong practitioner in Beijing received an e-mail from editor@minghui.ca, the legitimate address of one of the main Falungong sites in Canada. The message contained a fake Li Hongzhi article. Closer inspection of the e-mail header reveals that the message actually originated from IP address 202.106.227.134, which is located in Beijing, not Canada.[121] An additional harassing technique reported by Falungong practitioners abroad is e-mail bombardment. According to interviews, the personal e-mail of group members, as well as the official Falungong e-mail addresses, is regularly saturated with hundreds or thousands of bogus messages (as many as 20,000, in some cases), making it impossible to use the Internet for reliable communication. Similarly, Falungong web pages are repeatedly saturated with bogus requests from IP addresses that do not exist, preventing other users from visiting the pages.[122]

---

[119]"China Establishes Web Page to Combat 'Cults,'" *Xinhua,* May 10, 2000, in FBIS, May 10, 2000.

[120]Maureen Pao, "Information Warrior," *Far Eastern Economic Review,* February 4, 1999. See also "PRC Said Interfering in Information Center's Work," Hong Kong Information Center of Human Rights and Democratic Movement in China, January 8, 1999, in FBIS, January 8, 1999.

[121]The full header of this e-mail can be found in http://hrreport.buhuo.net/book2e/eb210-1.html#3.

[122]*New York Times,* September 9, 2000.

Members of the overseas dissident community, which has been riven by personality and ego clashes, appear united in their belief that the Chinese government is spreading rumors and planting false agents in their midst, with the goal of furthering dividing the already notoriously factionalized movement. In October 1999, Wei Jingsheng told a Taiwanese news agency that the Beijing government is using the Internet to spread rumors intended to "sow seeds of discord" among organizations of mainland Chinese dissidents.[123] Interviews in June 2000 with members of prominent dissident groups confirm these perceptions. If the Chinese government is indeed pursuing this strategy, it is succeeding beyond its wildest dreams. Dissident bulletin boards and chat rooms are regularly filled with accusations and counteraccusations, primarily revolving around the question of which dissidents are actually paid agents of the Ministry of State Security. It is generally impossible to ascertain conclusively the true identities of the posters of such messages, making it relatively easy for the Chinese security services to engage in clandestine disinformation operations of this type.

MEASURING SUCCESS

Numerous possible metrics exist for measuring the government's success in thwarting the political use of the Internet by domestic and external forces, but one salient fact stands out: There are currently no organizations inside or outside China using the Internet to mount a credible threat to the survival of the regime. Indeed, the 1999 State Department human-rights report asserted that by the end of 1998, there were no organizations posing a legitimate threat to the regime, regardless of their means of communication:

> By year's end [1998], almost all of the key leaders of the China Democracy Party (CDP) were serving long prison terms or were in custody without formal charges, and only a handful of dissidents nationwide dared to remain active publicly. Tens of thousands of members of the Falungong spiritual movement were detained after the movement was banned in July; several leaders of the movement were sentenced to long prison terms in late December and hun-

---

[123]"Internet Used to Promote Freedom of Expression in China," Taiwan Central News Agency, October 2, 1999.

dreds of others were sentenced administratively to reeducation through labor in the fall.[124]

The situation did not improve in 1999, 2000, or 2001. Indeed, by most measures, it has since deteriorated. As a recent State Department report on human rights in China observed, "only a handful of political dissidents remained active" by the end of 2000.[125]

FUTURE TRENDS

Does Beijing's relative success in controlling dissident use of the Internet indicate that the Internet will have only a limited effect on China's political landscape? Some analysts argue that, at least in the short to medium term, the spread of the Internet will tend to benefit authoritarian regimes at the expense of dissidents and pro-democracy activists. As Kalathil and Boas observe, for example, China and other authoritarian states have responded effectively to the dissident challenge by implementing a combination of reactive measures, including blocking web sites and jailing activists, and proactive policies, such as distributing propaganda online and offering e-government services.[126] This has enabled China to minimize the Internet's potential as an instrument for "subversive" activity while simultaneously strengthening the position of the regime. "The Chinese state," Kalathil and Boas write, "has shown that it can use the Internet to enhance the implementation of its own agenda."[127] Indeed, authoritative pronouncements in the state-controlled media have called for more-extensive use of the Internet in "guiding opinion." An August 9, 2000, *People's Daily* commentator's article, for example, called for the strengthening of "positive propaganda and influence on the Internet" to enable the government's "ideological and political work" to become "more inclusive and influential." The

---

[124]U.S. Department of State, Bureau of Democracy, Human Rights, and Labor, *China Country Report on Human Rights Practices, 1999*, February 25, 2000.

[125]U.S. Department of State, Bureau of Democracy, Human Rights, and Labor, *China Country Report on Human Rights Practices, 2000*, February 23, 2001, p. 2.

[126]Shanthi Kalathil and Taylor C. Boas, "The Internet and State Control in Authoritarian Regimes: China, Cuba, and the Counterrevolution," Carnegie Endowment Working Papers, No. 21, July 2001, pp. 1–10, 15–18.

[127]Ibid., p. 8.

commentator went on to note that official Internet sites "accurately and comprehensively publicize the Party line, strategies and policies, and consistently guide people with correct state opinion."[128] Chinese scholars have also recognized the Internet's potential as an effective tool for carrying out "political thought work" and disseminating propaganda.[129] China's proactive efforts to use the Internet to bolster regime power, however, have thus far produced only limited results. For example, although Beijing has actively promoted a "government online" plan, a recent survey of e-government initiatives around the world found that China ranked eighty-third out of 196 countries.[130]

Measures of a more reactive nature will thus continue to occupy a dominant place in Beijing's strategy for dealing with dissident use of the Internet. There is some evidence that Beijing's technical counter-measures are becoming increasingly sophisticated. In future efforts to limit what Beijing views as the pernicious side effects of the spread of the Internet in China, however, the authorities will likely try to combine low-tech and high-tech measures. This evolving approach is illustrated by Beijing's recent efforts to tighten its control over China's nearly 100,000 Internet cafes.[131] During a nationwide sweep that began in April 2001, the authorities shut down more than 17,000 Internet cafes for allowing access to pornographic or "subversive" web sites, according to *Wen Hui Bao*, an official Shanghai-based newspaper. Some 28,000 more Internet cafes were ordered to install software that prevents users from accessing forbidden web sites and

---

[128]"Vigorously Strengthen the Building of China's Internet Media," commentator's article, *Renmin ribao*, August 9, 2000, in FBIS, August 9, 2000.

[129]Xie Haiguang (ed.), *Hulianwang yu sixiang zhengzhi gongzuo gailun* [*Introduction to the Internet and Political Thought Work*], Shanghai: Fudan University Press, 2000.

[130]World Markets Research Centre and Brown University, "Global E-Government Survey," September 2001. The United States placed first, followed by Taiwan, Australia, Canada, the United Kingdom, Ireland, Israel, Singapore, Germany, Finland, and France. The sophisticated e-government program of Taiwan stands in sharp contrast to the PRC's "government online" project. For example, Taiwan President Chen Shui-bian has a weekly e-mail Internet newsletter, called *President A-bian's E-Paper*. See www.president.gov.tw/1_epaper.iod.html. President Chen has also engaged in live, online chat sessions with Internet users in Taiwan.

[131]Some observers have speculated that the crackdown may have had more to do with the desire of the authorities to collect registration fees from delinquent Internet cafe proprietors than concerns about what Internet users were doing at the cafes.

enables the establishments to monitor user activities.[132] While there is little information available about the software, other official media reports indicate that Internet cafes in major cities such as Xi'an and Chongqing have installed programs called "Internet Police" and "Internet Cafe Security Management System" to block access to banned content and make it easier for police to track users engaging in "subversive" activities online.[133] Beijing does not rely on software alone to control what users do in Internet cafes. Indeed, the authorities rely just as heavily on Internet cafe owners and employees to prevent visitors from viewing banned material on the Web. Because their interests lie in earning profits, not engaging in politics, the proprietors of many of the estimated 94,000 Internet cafes in China are willing to cooperate with local authorities.

While Beijing's countermeasures have been relatively successful on the whole, the current lack of credible challenges to the regime, despite the introduction of massive amounts of modern telecommunications infrastructure, does not lead inexorably to the conclusion that the regime will continue to be immune from the forces unleashed by the increasingly unfettered flow of information across its borders. While Beijing has done a remarkable job of finding effective counterstrategies to the potential negative effects of the information revolution, the scale of China's information-technology modernization would suggest that time is eventually on the side of the regime's opponents. Nina Hachigian predicts that "control over information will slowly shift from the state to networked citizens," leading to potentially "seismic" changes.[134] At least a few Chinese scholars are in agreement with the basic thrust of this assessment. In the words of one Chinese researcher, "It will be impossible to control this technology completely, even with filters and an army of trained digital agents."[135]

---

[132]"17,000 Internet Bars Shut Down," *South China Morning Post*, November 21, 2001.

[133]"Internet Police Software Installed in 800 Xi'an City Internet Bars," *Xinhua*, August 7, 2001; see also "New Software Censors Web in Chongqing Net Cafes," *China Online*, August 14, 2001, which cites other official media reports.

[134]Nina Hachigian, "China's Cyber-Strategy."

[135]Chinese researcher, January 2001.

In the short to medium term, however, the multifarious close part-
nerships between the regime and the commercial Internet sector,
institutionalized in a complex web of regulations and fiscal relationships, imply that the government will not lose the upper hand soon.
The government's strategy is also aided by the current economic environment in China, which encourages the commercialization of the
Internet, not its politicization. As one Internet executive put it, for
Chinese and foreign companies, "the point is to make profits, not political statements."[136] Thus, the Internet, despite the rhetoric of its
most enthusiastic supporters, will probably not bring "revolutionary"
political change to China, but instead will be a key pillar of China's
slower, evolutionary path toward increased pluralization and pos-
sibly even nascent democratization.

As dissident use of the Internet and regime countermeasures continue to evolve in China and other authoritarian countries, future research should not only update this unfolding story but should also
place it in a comparative framework to help enhance our under-
standing of the political impact of the Internet in these countries.
Comparing the political role of the Internet in a variety of authoritarian, quasi-authoritarian, and nondemocratic states in Asia—namely,
China, Myanmar, Singapore, Vietnam, and Malaysia—would represent a useful next step in this process. The research agenda could
then be broadened to compare the situation in these countries with
that in states outside the region, perhaps including Iran, Saudi Arabia, and Cuba, to produce more widely applicable conclusions about
the political use of the Internet in authoritarian countries.

---

[136]Interview, U.S. businessperson, 2001.

DISSIDENT WEB SITES

CHINESE DEMOCRACY MOVEMENT WEB SITES[1]

Organization	URL[2]
Beijing Spring	www.bjzc.org (C)
China Democracy Party	members.tripod.com/~chinadp/ (C)
China Democracy Party	http://www.freechina.net/cdp/ (E)
China Democracy Party	www.geocities.com/wellesley/gazebo/ 9797 (C)
China Democracy Information Network	www.dinfo.org

---

[1]The Internet is a highly fluid information environment, with new web sites constantly appearing and old ones constantly disappearing. It is difficult to keep pace with these changes, and in the case of the web sites of Chinese dissident organizations, it is even more challenging, for several reasons. Chief among them is that many groups change their web addresses frequently in order to thwart attempts by the authorities to block access to them. Web sites also become defunct when groups disband or no longer have the resources to maintain their online presence. When this list was compiled in January 2002, all of the links were working, unless otherwise noted. Of course, by the time of publication, it is possible that some links will have become outdated.

[2]Chinese-only web sites are identified here with a (C), English-only websites with an (E), German-only web sites with a (G), French-only web sites with an (F), and Tibetan-only web sites with a (T). The URLs of multilingual websites are followed with a notation such as (C,E), (C,E,T), or (C,E,F). Languages that are not commonly used on Chinese dissident web sites, such as Hindi, for example, are noted by their full names. In some cases, the English versions of multilingual web sites do not contain all of the information available in Chinese.

Index of Chinese Dissident Websites	http://www.chinasite.com/dissident.html
Chinese Democratic Justice Party	members.tripod.com/~chinadp/
China Labor Party/ China Workers Party	http://newstrolls.com/news/dev/guest/ 010899-2.htm (E) www.freechina.net/cwp/ (C) members.aol.com/wenxiao (C)
China Labour Bulletin	www.china-labour.org.hk (C,E)
China Spring	www.chinaspring.org (C)
China Monthly (*Minzhu Zhongguo*)	www.chinamz.org (C)
Democracy Forum (*Minzhu Yazhou Jijinhui*)	www.asiademo.org (C)
Free China Movement	www.freechina.net (C,E,F)
Freenet China	www.freenet-china.org (C)
Home for Global Internet Freedom	www.internetfreedom.org (C)
Hong Kong Information Center for Human Rights and the Democratic Movement in China	www.89-64.com (C) www.89-64.com/english/indexen.html (E)
Hong Kong Voice of Democracy	www.democracy.org.hk (E)
Human Rights in China	www.hrichina.org (C,E)
June4.org	www.june4.org (E)
Lin Hai	www.linhai.org (no longer available) www.freechina.com (E) home4u.china.com/technology/internet/ hopy/ (no longer available)
Modern Chinese Problem Research Forum	members.tripod.com/~China_Forum/ index.html (E)

Party for Freedom and Democracy in China	http://www.freechina.net/pfdc/ (E)
Press Freedom Guardian	www.pressfreedom.com (no longer available)
Reforming China Network	www.freechina.net/rcn/ (no loner available)
Ren Wanding's Human Rights Union	www.freechina.net/rwd/ (C)
Silicon Valley for Democracy in China	www.svdc.org (E)
Support Democracy in China	www.christusrex.org/www1/sdc/sdchome.html (E)
Tunnel	www.geocities.com/SiliconValley/Bay/5598 (C)
VIP Reference (Da Cankao)	www.bignews.org (C)
Zhang Weiguo's New Century Net	http://www.ncn.org/zwgInfo/index.asp (C,E)

OTHER DISSIDENT WEB SITES

Organization	URL	Comments
Tibet		
Official Tibetan Government-in-Exile Pages		
Tibetan Government in Exile, Dharamsala, India	www.tibet.com (E) www.tibet.net (E,C,T,and Hindi)	Official site of the Tibetan Government in Exile, maintained by Office of Tibet, London

Tibetan Government Department of Information and International Relations	tibetnews.com (E)	Site includes the *Tibetan Bulletin*, the official journal of the Tibetan exile government
Canada Tibet Committee	www.tibet.ca (E,F)	Main Tibetan advocacy site in Canada
Committee of 100 for Tibet	www.tibet.org/ Tibet100/ (E)	Organization dedicated to informing the public about Tibet and working for human rights
Dalai Lama Official Web Site	www.dalailama.com (E)	
Free Tibet Campaign	www.freetibet.org (E)	UK-based group
International Campaign for Tibet	www.savetibet.org (E)	Washington, D.C.-based group; numerous campaigns and appeals for action
Tibet Justice Center	www.tibetjustice.org/ (E)	Formerly International Committee of Lawyers for Tibet
Students for a Free Tibet	www.tibet.org/SFT/ (E)	Contains links to numerous SFT chapters in the U.S. and Canada
Swiss-Tibetan Friendship Association	www.tibetfocus.com (G,F)	Primarily German and French language

The Milarepa Fund	www.milarepa.org (E)	Sponsor of annual Tibetan Freedom Concerts
The Tibetan Plateau Project	www.earthisland.org/ tpp/ (E)	Tibetan environmental organization
Tibet Environmental Watch	www.tew.org (E)	Tibetan environmental organization
Tibet Fund	www.tibetfund.org (E)	N.Y.-based fundraising group
Tibet House	www.tibethouse.org (E)	Based in N.Y.
Tibet Information Network (TIN)	www.tibetinfo.net (E)	Information on current develop-ments in Tibet; publishes reports on social, eco-nomic, and political issues
Tibet Links	www.geocities.com/ Athens/Academy/ 9594/links.html (E)	Contains over 370 links to web sites with political, cultural, religious, and other material related to Tibet; also contains links to web sites of several Uyghur and Mongolian pro-independence groups

Tibet Online	www.tibet.org (E)	San Francisco area group that provides the Tibetan exile government with Internet assistance (also has links to 56 foreign-language pro-Tibet sites)
Tibetan Center for Human Rights and Democracy	www.tchrd.org (E)	NGO based in Dharamsala, India; site includes newsletter and e-mail bulletin is also available
Tibetlink	www.tibetlink.com (E)	Chat and BBS
U.S. Tibet Committee	www.ustibet.org (E)	
Voice of Tibet	www.vot.org (C,E,T)	Broadcasts in Tibetan
Worldbridges Tibet	Worldbridges.com/Tibet/ (E)	Includes chat room and BBS
World Tibet Day	www.worldtibetday.org (E)	Organizing and information site about annual World Tibet Day

Pro-Independence/Radical Tibetan Exile Groups

| International Tibet Independence Movement (ITIM) | www.rangzen.com (E) | Pro-independence group based in Indiana |

Tibetan Youth Congress	www.Tibetanyouth congress.org (E)	Pro-independence Tibetan exile youth organization; site contains *Rangzen Magazine* (*Independence Magazine*)

Uyghurs

East Turkestan National Freedom Center	www.uyghur.org (no longer available)	Based in Washington, D.C.; site last updated June 1999
Uygur Awazi Radiosi	www.uygur.com	Primarily songs and videos
Free Eastern Turkestan	www.taklamakan.org (E)	
Uyghur Human Rights Coalition	www.uyghurs.org (E)	Based in Washington, D.C.; includes several campaigns to release prisoners, etc.
Eastern Turkistan Information Center	www.uygur.org (E,G, and Uyghur)	Based in Munich, Germany
Citizens Against Chinese Communist Propaganda, Eastern Turkestan Information	www.caccp.org/et/ (E)	Info on CACCP's opposition to China theme park in Kissimmee, Florida

Mongols

Inner Mongolian People's Party	www.innermongolia.org (C,E, and Mongolian)	Includes info on campaigns to free imprisoned activists; also has bumper sticker and t-shirts for sale
Oyunbilig's Great Mongol Home Page	www.mongols.com (E)	

Falungong

Location/Name	URL
North America	
United States	www.falundafa.org (C,E,F, Vietnamese, Russian, Bulgarian, Spanish, and Dutch)
"Perfection" Site (U.S.)	www.esatclear.ie/~huiwu/ (C)
"Scientific Discoveries" Site (U.S.)	www.science-discover.net/index_c.htm (no longer available)
"Witness" Site (U.S.)	Falunwitness.net (no longer available)
"Wisdom" Site (U.S.)	Falunwisdom.net (no longer available)
Falungong in North America	minghui.ca (C,E, and many other languages)
Canada	falundafa.ca (C,E,F, and many other languages)
Ottawa, Canada	www.falundafaottawa.cjb.net/ (C,E)
Quebec City, Canada	Pages.infinit.net/lucbm/falundafa-qc/ (F)
Simon Fraser University, British Columbia	www.sfu.ca/~falun/ (E)
University of Toronto	www.campuslife.utoronto.ca/groups/falun/ (E)

| Worldwide Falundafa Day | www.worldfalundafaday.org (C,E) |

Europe

Austria and Germany	www.falundafa.de (G)
Denmark	www.falundafa.dk (Dutch)
Europe	www.falundafa.nu/ (C,E)
Ireland	www.falundafa.nu/eire/ (E)
Norway	hem.fyristorg.com/falundafa.se/Norge/ NORGE.HTM (Norweigian)
Slovakia	www.falungong.sk (Slovak)
Sweden	www.falungong.net/sweden (Swedish)
Switzerland	www.falundafa.ch/ (G,F, and Italian)
St. Petersburg, Russia	Users.nevalink.ru/falundafa (Russian)

Asia

Australia	Falundafa.au.cd/ (E)
Beijing, China	www.beijingnet.com/home/falun/falun.htm (shut down by Chinese authorities)
Guangzhou, China	Scut.edu.cn (shut down by Chinese authorities)
Daqing, China	(shut down by Chinese authorities)
Hong Kong	Home.netvigator.com/~falunhk (C,E)
Japan	www.falundafa-jp.net (Japanese)
Korea	www.falundafa.or.kr (Korean)
Malaysia	http://61.6.32.133/falundafa/ (C,E)
New Zealand	Homepages.ihug.co.nz/~maxs/falundafa. html (E)
Singapore	Web.singnet.com.sg/~falun (C,E)
Taiwan	www.falundafa.org.tw (C,E)

Middle East

Israel   www.falundafa.org.il (C,E, Russian, Hebrew, and Arabic)

REFERENCES

Barme, Geremie, and Sang Ye, "The Great Firewall of China," *Wired 5.06*, June 1997.

Becker, Jasper, "Review of Dissidents, Human Rights Issues," *South China Morning Post*, January 12, 1999.

Booz Allen & Hamilton, *E-Commerce at the Grass Roots: Implications of a "Wired" Citizenry in Developing Nations*, prepared for the National Intelligence Council, June 30, 2000.

Buruma, Ian, *Bad Elements: Chinese Rebels from Los Angeles to Beijing*, Random House, 2001.

Cameron, Maxwell A., Robert J. Lawson, and Brian W. Tomlin (eds.), *To Walk Without Fear: The Global Movement to Ban Landmines*, New York: Oxford University Press, 1999.

Cao, Xueyi, "Here Comes the Wolf, Raise Your Hunting Rifle—Be Alert to Computer Network Security," *Jiefangjun bao*, August 25, 1999, p. 5, in FBIS, August 25, 1999.

Chen, Chiu, "University Students Transmit Messages on Defending the Diaoyu Islands Through the Internet, and the Authorities Are Shocked at This and Order the Strengthening of Control," *Sing Tao Jih Pao*, September 17, 1996, in FBIS September 18, 1996.

Chen, Jian, "*Zhongguo hulianwang xingye zilu gongyue zhengshi qianshu*" [Chinese Internet Industry Self-Discipline Pact Formally Signed], *Renminwang*, March 26, 2002.

Chen, Ting, and He Jing, "Pay Attention to Phenomenon of 'Information Colonialism,'" *Jiefangjun bao*, February 8, 2000.

"China: Activists Launch Online Magazine," Reuters, June 18, 1997.

"China Charges Dissident Author with Subversion," Associated Press, December 22, 1999.

"China Charges Student on Falungong E-mail," Reuters, November 8, 1999.

"China Cyber-Cops Partially Block Hacked Web Site," Reuters, October 29, 1998.

China Democracy Party, "New Century Declaration," December 31, 1999.

_____, "Open Declaration on the Establishment of the China Democracy Party Zhejiang Preparatory Committee" (*Zhongguo minzhudang Zhejiang choubei weiyuanhui chengli gongkai xuanyan*).

_____, "Political Program of the China Democracy Party" (*Zhongguo minzhudang zhengzhi gangling*).

"China Enforces Control of Electronic Public Notice Service over the Internet," *Xinhua*, November 6, 2000.

"China Establishes Web Page to Combat 'Cults,'" *Xinhua*, May 10, 2000, in FBIS, May 10, 2000.

"China Grants Early Release to Cyberdissident," Associated Press, March 3, 2000.

"China: Information Security," Report from U.S. embassy, Beijing, June 1999.

China Internet Network Information Center (CNNIC), "Survey Statistical Report on the Development of the Chinese Internet" (*Zhongguo hulian wangluo fazhan zhuangkuang tongji baogao*), January 2002.

_____, "Semiannual Survey Report on the Development of China's Internet," July 2001.

_____, "Semiannual Survey Report on the Development of China's Internet," January 2001.

_____, "Semiannual Survey Report on the Development of China's Internet," July 2000.

_____, "Semiannual Survey Report on Internet Development in China," January 2000.

_____, "Semiannual Survey Report on Internet Development in China," July 1999.

_____, "Statistical Report of the Development of China Internet," January 1999.

_____, "Survey Statistical Report on the Development of the Chinese Internet" (*Zhongguo* Internet *fazhan zhuangkuang tongji baogao*), July 1998.

_____, "Survey Statistical Report on the Development of the Chinese Internet" (*Zhongguo* Internet *fazhan zhuangkuang tongji baogao*), October 1997.

"China Sets New Net Rules," *South China Morning Post*, January 21, 2002.

"China Tightens Grip on Booming Net Cafes," *South China Morning Post*, July 30, 2001.

"China: We're Only Human," Reuters, October 26, 1998.

"China's Internet Industry Wants Self-Discipline," *People's Daily Online*, 26 March 2002.

"China's Internet Information Skirmish," Report from U.S. embassy Beijing, January 2000.

"Chinese Intellectual Detained for Alleged Internet Crimes," *Inside China Today*, www.insidechina.com/news, September 6, 1999.

"Chinese Internet Writer Faces Trial for Subverting State Power," e-mail press release from *VIP Reference* editor Richard Long, May 22, 2000.

"Chinese Organ Screens Web Site," *Far Eastern Economic Review*, September 21, 2001.

"Circular Issued on Destroying Falungong Publications," *Xinhua*, July 28, 1999, in FBIS, July 28, 1999.

Cisneros, Oscar, "ISPs Accuse China of Infowar," *Wired News*, July 30, 1999.

Clarke, Ian, Scott G. Miller, Theodore W. Wong, Oskar Sanderg, and Brandon Wiley, "Protecting Free Expression Online with Freenet," *IEEE Internet Computing*, January-February 2002, pp. 40-49.

"Court Verdict on Dissident Lin Hai," Hong Kong Information Center, January 20, 1999.

Danitz, Tiffany, and Warren P. Strobel, "Networking Dissent: Cyber Activists Use the Internet to Promote Democracy in Burma," in John Arquilla and David Ronfeldt (eds.), *Networks and Netwars*, Santa Monica, CA: RAND, 2001, pp. 129–170.

Digital Freedom Network, "Attacks on the Internet in China: Chinese Individuals Currently Detained for Online Political or Religious Activity," available on the Digital Freedom Network website at http://www.dfn.org/focus/china/netattack.htm.

Digital Freedom Network, "Attacks on the Internet in China: Internet-Related Legal Actions and Site Shutdowns Since January 2000," available on the Digital Freedom Network website at http://www.dfn.org/focus/china/shutdown.htm.

"Dissidents to Start Signature Campaign," Hong Kong Information Center, May 20, 1999, in FBIS, May 20, 1999.

Dobson, William J., "Dissidence in Cyberspace Worries Beijing," *San Jose Mercury News*, June 28, 1998.

"Doctor Jailed for Promoting Falungong on Internet" http://www.june4.org./news/database/jan2000/doctorjailed.html.

Dorgan, Michael, "Chinese Censors Losing Online Race," *San Jose Mercury News*, May 22, 2000.

_____, "Critics of Taiwan Policy Outwit the Censor," *South China Morning Post*, May 25, 2000.

Downs, Erica Strecker, and Phillip C. Saunders, "Legitimacy and the Limits of Nationalism: China and the Diaoyu Islands," *International Security*, Vol. 23, No. 3, Winter 1998/99, pp. 114–146.

Dube, Jonathan, "China Ate My Web Site," ABCNEWS.com, August 6, 1999.

Eckholm, Erik, "China Sect Members Covertly Meet Press and Ask World's Help," *New York Times*, October 29, 1999.

_____, "China Cracks Down on Dissent in Cyberspace," *New York Times*, December 31, 1997.

"Eight Reactionary Activists from Jilin Who Were Imprisoned Before Call for Reversing the Verdict on the June 4th Incident and for Government Compensation," Hong Kong Information Center for Human Rights and Democracy, June 3, 2000, in FBIS, June 3, 2000.

"Ensuring PRC Military Network Security," *Xiandai junshi*, October 11, 1999, pp. 35–36.

Falungong, "A Report on Extensive and Severe Human Rights Violations in the Suppression of Falungong in the People's Republic of China, 1999–2000."

"Falungong and the Internet: A Marriage Made in Web Heaven," VirtualChina.com, July 30, 1999.

"Falungong Hot on Jiang's Trail," *Agence France Presse*, September 7, 1999.

Farley, Maggie, "Hactivists Besiege China," *Los Angeles Times*, January 4, 1999.

Forney, Matt, "Dissonant Dissent," *Far Eastern Economic Review*, June 5, 1997, pp. 28–32.

Foster, William, and Seymour E. Goodman, *The Diffusion of the Internet in China*, Center for Security and Cooperation, Stanford, CA: Stanford University, November 2000.

"Four CDP Founders Given Stiff Prison Sentences," Hong Kong Information Center for Human Rights and Democracy, November 9, 1999, in FBIS, November 9, 1999.

Franda, Marcus, *Launching into Cyberspace: Internet Development and Politics in Five World Regions*, Lynne Reinner, 2001.

Global Petition Campaign/June4.org, "Wang Dan Demands Change in China," press release, undated.

Goodspeed, Peter, "Falung Gong, Beijing Wage War over Internet," *National Post*, November 2, 1999.

Gries, Peter Hays, "Tears of Rage: Chinese Nationalist Reactions to the Belgrade Embassy Bombing," *The China Journal*, No. 46, July 2001, pp. 25–43.

"*Guanyu Ziyouwang*" [About Freenet], December 11, 2001, available at www.internetfreedom.org/gb/articles/1042.html.

"Guizhou Poet Ma Zhe Has Been Sentenced to Five Years' Imprisonment on Subversion Charge," Hong Kong Information Center, March 14, 2000.

Guo, Liang, and Bu Wei, *Investigative Report on Internet Use and Its Impact (Hulianwang shiyong zhuangkuang ji yingxiang de diaocha baogao)*, Chinese Academy of Social Sciences, Center for Social Development, research supported by State Informatization Office, April 2001.

_____, "The Questionnaire and Responses to a Survey on Internet Usage and Impact in Beijing, Shanghai, Guangzhou, Chengdu, and Changsha," Chinese Academy of Social Sciences, Center for Social Development, 2001.

Hachigian, Nina, "China's Cyber-Strategy," *Foreign Affairs*, Vol. 80, No. 2, March/April 2001, pp. 118–133.

"Hacker Attacks Society for Human Rights Studies Web Site," *Xinhua*, October 29, 1998, in FBIS, October 29, 1998.

"Hangzhou Court Verdict on Wang Youcai," Hong Kong Information Center, December 21, 1998, in FBIS, December 21, 1998.

"Hangzhou Security Bureau Detains Five More Dissidents," Hong Kong Information Center of Human Rights and Democratic Movement in China, June 19, 1999.

Harwit, Eric, and Duncan Clark, "Shaping the Internet in China: Evolution of Political Control over Network Infrastructure and Content," *Asian Survey*, Vol. 41, No. 3, May/June 2001, pp. 377–408.

Hoh, Erling, "Freedom's Factions," *Far Eastern Economic Review*, March 4, 1999, pp. 26–27.

Hu, Jim, "Ten Years Later: Chinese Dissidents Using Net," CNET News.com, June 8, 1999.

Hua, Chen, "PLA Spy Major General Liu Liankun Has Extensive Interpersonal Relationships and Strong Backing and His Execution Was Enforced by the Ministry of State Security," *Ming Pao*, September 19, 1999, p. A12, in FBIS, September 19, 1999.

Human Rights in China, "Tiananmen Mothers Seed Global Support for Campaign to End Impunity: On-Line Petition Launched," press release, June 2, 2000.

"Internet Allows Chinese Dissidents to Network," www.nando.net, June 2, 1998.

"Internet Police Software Installed in 800 Xi'an City Internet Bars," *Xinhua*, August 7, 2001.

"Internet Used to Promote Freedom of Expression in China," Taiwan Central News Agency, October 2, 1999.

"Jiang Stresses Stability, Unity," *Xinhua*, December 18, 1998, in FBIS, December 18, 1998

"Jiang Zemin Orders State and Military Security Departments Guard Against the Leaking of State Secrets Via the Internet," *Ming Pao*, September 22, 1998, p. A14, in FBIS, September 22, 1998.

"Jiang Zemin Stresses Rural Stability at NPC Panel Discussions," *Xinhua*, March 5, 2002, in FBIS, March 5, 2002.

"Jiang, Zhu Speak at Politics and Law Conference," *Xinhua*, December 23, 1998, in FBIS, December 24, 1998.

Kalathil, Shanthi, and Taylor C. Boas, "The Internet and State Control in Authoritarian Regimes: China, Cuba, and the Counterrevolution," Carnegie Endowment Working Papers, No. 21, July 2001.

Kennedy, Andy, "For China, the Tighter the Grip, the Weaker the Hand," *Washington Post*, January 17, 1999.

"Kids, Cadres, and 'Cultists' All Love It: Growing Influence of the Internet in China," report from U.S. embassy, Beijing, March 2001.

Lakshmi, Rama, "Young Lama Inspires Tibetan Exile Youth," *Washington Post*, May 29, 2000.

Lam, Willy Wo-Lap, "Beijing Orders Close Watch on 150 Dissidents," *South China Morning Post*, December 24, 1998, p. 1.

_____, "Big Push to Maintain Stability in 1999," *South China Morning Post*, January 5, 1999.

_____, "China Development Union Leaders Arrested, Told to Close," *South China Morning Post*, October 27, 1998, p. 9.

_____, "Falungong Protest Shocks Party Leadership," *South China Morning Post*, April 28, 1999.

Langfitt, Frank, "Taking Dissent Online in China; E-mail: In the Age of the Internet, Chinese Leaders Are Finding It Harder to Contain Free Speech," *The Baltimore Sun*, May 11, 1999, p. 2A.

Lee, Jennifer 8., "Punching Holes in Internet Walls," *New York Times*, April 26, 2001.

_____, "U.S. May Help Chinese Evade Net Censorship," *New York Times*, August 30, 2001.

Liu, Melinda, "The Great Firewall of China," *Newsweek International*, October 11, 1999.

Liu, Youshui, and Zhang Wusong, "High-Tech Development and State Security," *Jiefangjun bao*, January 11, 2000, p. 6, in FBIS, January 11, 2000.

Ma, Josephine, "Cyber-Crackdown Fails to Silence Protestors," *South China Morning Post*, June 2, 2000.

_____, "Defiant Cyber-Surfers Play Cat-and-Mouse Game," *South China Morning Post*, June 8, 2000.

_____, "Police Charge Web Site's Founder with Subversion," *South China Morning Post*, June 8, 2000.

Martin, David, and Andrew Schulman, "Deanonymizing Users of the Safeweb Anonymizing Service," February 11, 2002, http://www.cs.deanonymizing-safeweb.pdf.

Mufson, Steve, "Chinese Protest Finds a Path on the Internet," *Washington Post*, September 17, 1996, p. A9.

"New Party for Workers to Seek Registry," Hong Kong *Agence France Presse*, January 2, 1999, in FBIS, January 2, 1999.

"New Software Censors Web in Chongqing Net Cafes," *China Online*, August 14, 2001.

"Non-Governmental Report Center Web Page Set Up," Hong Kong Information Center for Human Rights and Democracy, April 3, 1999.

Oram, Andy (ed.), *Peer-to-Peer: Harnessing the Power of Disruptive Technologies*, O'Reilly, 2001.

Pao, Maureen, "Information Warrior," *Far Eastern Economic Review*, February 4, 1999, pp. 26–27.

Platt, Kevin, "China's 'Cybercops' Clamp Down; Beijing Sees Growing Web Use as Threat, But It Had a Victory Nov. 9 in Connection with Four Convictions," *Christian Science Monitor*, November 17, 1999, p. 6.

"Police Arrest Dissidents to Prevent Seminar Opening," Hong Kong Information Center for Human Rights and Democratic Movement in China, March 14, 1999, in FBIS, March 14, 1999.

"Police Claim Falungong Followers Leaked State Secrets," *Xinhua*, October 25, 1999, in FBIS, October 25, 1999.

Pomfret, John, "With Carrots and Sticks, China Quiets Protesters," *Washington Post*, March 22, 2002.

_____, "China Cracks Down on Worker Protests," *Washington Post*, March 21, 2002.

_____, "Thousands of Workers Protest in Chinese City," *Washington Post*, March 20, 2002.

_____, "China Sect Penetrated Military and Police: Security Infiltration Spurred Crackdown," *Washington Post*, August 7, 1999.

"PRC Internet Police Said in On-Line Conflict with China Democracy Party," *Tai Yang Pao*, April 23, 2000, in FBIS, April 24, 2000.

"PRC Net Dreams: Is Control Possible?" report from U.S. embassy, Beijing, September 1997.

"PRC Said Interfering in Information Center's Work," Hong Kong Information Center of Human Rights and Democratic Movement in China, January 8, 1999, in FBIS, January 8, 1999.

"PRC Web Forums on Mid Air Collision," report from U.S. embassy, Beijing, April 2001, www.usembassy-china.org.cn/english/sandt/midaircollison-webforums.html.

"PRC's Cyber-Dissident Released from Jail Early," Hong Kong *Agence France Presse*, March 3, 2000.

"Pro-Democracy Activist Interviewed on Party Formation," *Mainichi Shimbun*, September 13, 1998, in FBIS, September 13, 1998.

"Proxy Servers," *China Rights Forum*, Human Rights in China, fall 1998.

"Report on PRC Controlling Dissidents' E-mail," Hong Kong Information Centre for Human Rights and Democratic Movement in China, January 19, 2000, in FBIS, January 19, 2000.

Ronfeldt, David, John Arquilla, Graham E. Fuller, and Melissa Fuller, *The Zapatista Social Netwar in Mexico*, Santa Monica, CA: RAND, 1998.

Savitt, Scott, "China's Internet Revolution," *Asian Wall Street Journal*, December 21, 1999, p. 10.

Schmit, Julie, and Paul Wiseman, "Surfing the Dragon: Web Surfers Find Cracks in Wall of Official China," *USA Today*, March 15, 2000, p. 01B.

Sheff, David, *China Dawn: The Story of a Technology and Business Revolution*, New York: Harper Business, 2002.

"*Shenme shi* PGP?" [What is PGP?], November 12, 2001, available at http://www.internetfreedom.org/gb/articles/1034.html.

Shi Lei, "*Xinxi bailinqiang: tupo zhonggong wangluo dianzi youjian fengsuo (zhiyi)*" [The Information Berlin Wall: Breaking the Chinese Communist Party's Net and E-Mail Blockade (Part One)], November 2, 2001, available at http://internetfreedom.org/gb/articles/994.html.

_____, "*Zhonggong ruhe guolu he jiecha wangluo dianzi youjian: tupo zhonggong wangluo dianzi youjian fengsuo (zhi er)*" [How the Chinese Communist Party Filters and Monitors the Net and E-Mail (Part Two)], November 2, 2001, available at http://www.internet freedom.org/gb/articles/997.html.

"Six Falungong Academics Jailed," *South China Morning Post*, December 24, 2001.

Smith, Craig S., "Sect Clings to the Web in the Face of Beijing's Ban," *New York Times*, July 5, 2001

"Some Special Notes for Network Users in CN Domain (Mainland China) or Accessing to the Internet from Behind Firewall (Need a Proxy?)," www.cnd.org.

"State Tracks Dissidents Online," Associated Press, March 24, 2000.

Stellin, Susan, "Terror's Confounding Online Trail," *New York Times*, March 28, 2002.

"Student Net Site Closed over Talk of Tiananmen," Reuters, September 6, 2001.

Sun, Wen, "Use Computer to Fight Crime and Pornography," *Renmin gongan bao*, February 8, 1996, p. 3, in FBIS, February 8, 1996.

Svensson, Peter, "China Sect Claims Sites Under Attack," *Washington Post*, July 30, 1999.

"Talks Given by Officials of the State Council and the Chinese Communist Party Central Letters and Visit Bureaus," *Xinhua*, June 14, 1999, in FBIS, June 14, 1999.

Teng, Yue, "China Should Handle Information Security Independently," *Wen wei po*, July 12, 1999, p. A7, in FBIS, July 27, 1999.

"Top U.S. Institute Won't Bow to Dictatorship," Central News Agency, February 20, 2000.

"Two Democracy Party Members Detained," *Agence France Presse*, June 29, 1999.

United States Department of State, Bureau of Democracy, Human Rights, and Labor, *China Country Report on Human Rights Practices, 2001*, March 2002.

_____, *China Country Report on Human Rights Practices, 2000*, February 2001.

_____, *China Country Report on Human Rights Practices, 1999*, February 2000.

Usdin, Steve, "China Online," *Yahoo Internet Life*, January 1997.

"Using Legal Means to Guarantee and Promote Sound Development of Information Network," commentator's article, *People's Daily*, July 12, 2001, in FBIS July 12, 2001.

"Veteran Dissident Qin Yongmin Detained Again," *Agence France Presse*, October 27, 1998.

"Vigorously Strengthen the Building of China's Internet Media," commentator's article, *Renmin ribao*, August 9, 2000, in FBIS, August 9, 2000.

"*Wangba shangwang de yixie anquan wenti*" [Some Security Problems of Going On-Line at Internet Cafes], internetfreedom.org/gb/articles/979.html.

"Web Discussion Sample: True Democracy, Fake Democracy, or No Democracy?" Report from U.S. embassy Beijing, undated, www.usembassy-china.org.cn/english/sandt/webdemocracy.html.

"Web Sites of Falungong Hit," *Agence Prance Presse*, April 14, 2000.

"Wife Demands Return of Confiscated Items," *Agence Prance Presse*, February 18, 1999.

Wong, Bobson, "Chinese Democracy Activist Sentenced," Digital Freedom Network electronic newsletter, December 18, 2001.

World Markets Research Centre and Brown University, "Global E-Government Survey," September 2001.

Xie, Haiguang (ed.), *Hulianwang yu sixiang zhengzhi gongzuo gailun* [*Introduction to the Internet and Political Thought Work*], Shanghai: Fudan University Press, 2000.

Xu, Xiaofang, and Dan Aidong, "Serious Challenge to Information Network Security," *Jiefangjun bao*, July 20, 1999, p. 6, in FBIS, July 20, 1999.

Yang, Guobin, "The Impact of the Internet on Civil Society in China: A Preliminary Assessment" (forthcoming).

Yatsko, Pamela, "China's Web Censors Win One—For Now," *Fortune*, December 24, 2001.

"*Yong jiami fangshi anquan shiyong dianzi youjian*" [Using Encryption for E-Mail Security], www.internetfreedom.org/gb/articles/987. html.

"*Youguan jiamifa, yinshenshu, ji yinxieshu ziyuan*" [Resources on Cryptography, Anonymity, and Steganography], November 12, 2002, available at http://www.internetfreedom.org/gb/articles/1035.html.

Zhang, Weiguo, "Evading State Censorship," *China Rights Forum*, Human Rights in China, fall 1998.

Zhao, Ying, "Information Security Issues," *Jingji guanli*, No.5, May 5, 1998, pp. 16–17.

"*Zhongguo dui wangluo de zhuyao jiankong fangfa he duice*" [China's Main Methods of Supervising and Controlling the Net and Countermeasures], www.internetfreedom.org/gb/articles/1012. html.

"*Zhongguo hulianwang xingye zilu gongyue*" [Chinese Internet Industry Self-Discipline Pact], *Renminwang*, March 27, 2002, available at http://www.people.com.cn/GB/it/49/149/20020327/695927.html.

"Zhu Rongji: Reform Requires Social Stability, State Security," *Xinhua*, March 5, 2000.